"I am recommending this book because I don't think there is any that delves into this subject of the daily struggles and achievements of 'gender experts' working in environment and development organizations. As a 'gender expert' myself, I think it is indeed very timely for such a collection to reveal how gender experts navigate through the challenges and professional cultures within development organisations, as 'change agents' to bring about the transformation they are mandated to. Fascinating and intriguing – a much needed book to understand the workings within development organisations."

Chanda Gurung Goodrich, PhD, Senior Gender Specialist – Gender Lead, International Centre for Integrated Mountain Development (ICIMOD)

"Feminist movement faces new 'second generation' issues as it enters the third decade of the 21st century. Its three palpable seismic fault lines are: polarized North-South contexts, activists versus agency 'femocrats', and women professionals versus gender experts. Using polyvocal dialogues with representatives in these tectonic zones rather than monochrome essays, Resurrección and Elmhirst blend feminism's political economy, science technology studies in ecology and field activism to throw a powerful spotlight on the politicized pathways ahead. This is a book that must be read by professionals and gender experts thrown into uneasy company."

Dipak Gyawali, former Nepali minister of water resources, leading IDRC-supported research on water-induced disasters, gender and climate change

"This timely book reads like an act of solidarity, as it reviews debates around the struggles, compromises, and achievements of 'gender experts', and acknowledges their persistent efforts to disregard professional, personal and disciplinary boundaries. And it feels like a call to action, by reminding us that making gender justice central to addressing existential environmental threats must overcome all boundaries."

Ines Smyth, Independent Consultant, former Senior Gender Advisor, Oxfam

"At long last, a book we have been waiting for! Going past binaries of good feminists and co-opted femocrats, speaking to us directly through voices of 'gender-experts' and providing a sharp and reflective framework to think through their narratives, this is a book that speaks to all interested in bringing forth a more just and sustainable world. It highlights the work of the workers on the sustainability frontline, who everyday work across an epistemological interface, take charge and confront the vicissitudes of integrating, translating, embodying gender into environmental and development work, or in fact 'genderising a log-frame'."

Seema Arora-Jonsson, Professor of Rural Development, Swedish University of Agricultural Sciences

"What happens when a marine biologist, a mechanical engineer, and a water engineer encounter gender expertise? In this exciting new book, Bernadette Resurrección and Rebecca Elmhirst help us understand how gender experts negotiate the tensions that arise from a clash of very different kinds of knowledge, that is positivist science and feminism. Effectively expanding cutting-edge feminist literatures in political ecology and science and technology studies, they shed long overdue light on mainstreaming gender into various areas of environment and development. The book is a smart exercise in feminist reflexivity and offers a stimulating firework of ideas of what happens when gender experts work with scientists on issues ranging from climate change to water insecurity, and when they insert themselves in technologies ranging from geospatial imaging to modelling futures. Highly recommended!"

Elisabeth Prügl, Professor of International Relations, and Director, Gender Centre, Graduate Institute, Geneva

"Highlighting the politics of gender, knowledge, and expertise among practitioners this volume addresses a key question—how do practitioners and professionals navigate gender-related challenges in their everyday work, and what might this suggest about the broader politics of knowledge and expertise? From fields as diverse as agrobiodiversity, disaster risk, or marine biology, exploring these dynamics represents a novel and welcome addition to ongoing debates in feminist political ecology, science studies, and the politics of environment and development. Sharing these experiences also has the potential to inspire and support those working to realize a more equitable and sustainable world."

Leila M. Harris, Professor, University of British Columbia/ Institute for Gender, Race, Sexuality and Social Justice, Institute for Resources, Environment and Sustainability

"This book takes on the challenge of putting research into practice with curiosity and rigour. Through an exploration of the 'gender expert', the book shows the challenges of implementing conclusions from research. Gender and intersectional inequalities are well documented in development research, leading to the proliferation of 'gender experts' to address them. But such efforts encounter numerous obstacles at all levels of program design and implementation. By engaging the professional experiences of gender experts across a range of natural resource sectors, this book shows the importance of on-going dialogue between research and practice."

Andrea Nightingale, Professor, Department of Sociology and Human Geography, University of Oslo

"Understanding how change happens in environmental policy and practice in development organizations could scarcely be more urgent today. This book provides a fascinating and critical view into the world of 'gender experts', those who are simultaneously charged with transforming their institutions, while also coming up with policy 'asks', win-wins, technical fixes, and killer facts. By reflecting on their dilemmas, strategies and compromises, this book provides the inspiration and the insights that feminists need to continue chipping away at for change 'from within'."

Laura Turquet, Policy Advisor, UN Women and co-founder of the UN Feminist Network

"This is a stunning and original work by Bernadette Resurrección, Rebecca Elmhirst and their diverse writing collaborators, exploring the performance and organisational milieu confronting gender professionals in the pursuit of gender and environmental transformations. The gender experts' fields of knowledge may clash with those of their professional colleagues; they themselves are often learning on the job but are expected to deliver on outsized work expectations, creating cross-cutting gender platforms and also mainstreaming gender into internal silos of environment and development work. How is it working out? Read this book and learn!"

Meryl Williams, Chair, Gender in Aquaculture and Fisheries Section, Asian Fisheries Society

"It is with great pleasure I endorse this book. Through a series of conversations with 'gender experts', the two eminent scholars in the field of political ecology, Bernadette Resurrección and Rebecca Elmhirst, have successfully taken on the innovative task to unravel their daily struggles and achievements in environment and development organisations."

Ragnhild Lund, Professor Emeritus, Department of Geography, Norwegian University of Technology (NTNU)

Negotiating Gender Expertise in Environment and Development

This book casts a light on the daily struggles and achievements of 'gender experts' working in environment and development organisations, where they are charged with advancing gender equality and social equity and aligning this with visions of sustainable development.

Developed through a series of conversations convened by the book's editors with leading practitioners from research, advocacy and donor organisations, this text explores the ways gender professionals – specialists and experts, researchers, organizational focal points – deal with personal, power-laden realities associated with navigating gender in everyday practice. In turn, wider questions of epistemology and hierarchies of situated knowledges are examined, where gender analysis is brought into fields defined as largely techno-scientific, positivist and managerialist. Drawing on insights from feminist political ecology and feminist science, technology and society studies, the authors and their collaborators reveal and reflect upon strategies that serve to mute epistemological boundaries and enable small changes to be carved out that on occasions open up promising and alternative pathways for an equitable future.

This book will be of great relevance to scholars and practitioners with an interest in environment and development, science and technology, and gender and women's studies more broadly.

Bernadette P. Resurrección is Associate Professor in the Department of Global Development Studies at Queen's University, Kingston, Ontario, Canada, and formerly was a Senior Research Fellow at Stockholm Environment Institute (SEI).

Rebecca Elmhirst is Professor of Human Geography in the School of Environment and Technology at the University of Brighton, UK.

Routledge Studies in Gender and Environments

With the European Union, United Nations, UN Framework Convention on Climate Change, and national governments and businesses at least ostensibly paying more attention to gender, including as it relates to environments, there is more need than ever for existing and future scholars, policy makers, and environmental professionals to understand and be able to apply these concepts to work towards greater gender equality in and for a sustainable world.

Comprising edited collections, monographs and textbooks, this new *Routledge Studies in Gender and Environments* series will incorporate sophisticated critiques and theorisations, including engaging with the full range of masculinities and femininities, intersectionality, and LBGTIQ perspectives. The concept of 'environment' will also be drawn broadly to recognise how built, social and natural environments intersect with and influence each other. Contributions will also be sought from global regions and contexts which are not yet well represented in gender and environments literature, in particular Russia, the Middle East, and China, as well as other East Asian countries such as Japan and Korea.

Series Editor: *Professor Susan Buckingham, an independent researcher, consultant and writer on gender and environment related issues.*

Titles in this series include:

Gendered Approaches to Spatial Development in Europe
Perspectives, Similarities, Differences
Edited by Barbara Zibell, Doris Damyanovic and Ulrike Sturm

Gender and Religion in the City
Women, Urban Planning and Spirituality
Edited by Clara Greed

Transecology
Transgender Narratives on Environment and Nature
Edited by Douglas A. Vakoch

Negotiating Gender Expertise in Environment and Development
Voices from Feminist Political Ecology
Edited by Bernadette P. Resurrección and Rebecca Elmhirst

Negotiating Gender Expertise in Environment and Development

Voices from Feminist Political Ecology

Edited by
Bernadette P. Resurrección and Rebecca Elmhirst

Routledge
Taylor & Francis Group
LONDON AND NEW YORK

earthscan
from Routledge

SEI Stockholm Environment Institute

WEGO ITN

CGIAR GENDER Platform

First published 2021
by Routledge
2 Park Square, Milton Park, Abingdon, Oxon OX14 4RN

and by Routledge
52 Vanderbilt Avenue, New York, NY 10017

Routledge is an imprint of the Taylor & Francis Group, an informa business

British Library Cataloguing-in-Publication Data
A catalogue record for this book is available from the British Library

Library of Congress Cataloging-in-Publication Data
Names: Resurreccion, Bernadette P., editor. | Elmhirst, Rebecca, editor.
Title: Negotiating gender expertise in environment and development: voices from feminist political ecology / edited by Bernadette P. Resurrección and Rebecca Elmhirst.
Description: Abingdon, Oxon; New York, NY: Routledge, 2021. | Series: Routledge studies in gender and environments | Includes bibliographical references and index.
Identifiers: LCCN 2020028643 | ISBN 9780815386124 (hardback) | ISBN 9781351175180 (ebook)
Subjects: LCSH: Ecofeminism. | Women and the environment.
Classification: LCC HQ1194 .N44 2021 | DDC 304.2082–dc23
LC record available at https://lccn.loc.gov/2020028643

ISBN: 978-0-8153-8612-4 (hbk)
ISBN: 978-1-351-17518-0 (ebk)

Typeset in Times New Roman
by Deanta Global Publishing Services, Chennai, India

This book is dedicated to the memory of Victor Tsang, a dedicated father, husband and committed feminist.

Contents

Development Goals (SDGs), committed to by all countries in 2015, was supposed to integrate environment and development more fully than ever before in an agenda linking the '5 Ps' of people, planet, prosperity, peace and partnership, while 'leaving no-one behind'. But in this vital 'decade of action' leading up to the global goals' 2030 deadline, it is clear that much more needs to be done. It is not just that many of the 17 goals and 169 targets are off-track in many countries, but more significantly, that much of the original integrated, transformative vision has been lost, replaced by bureaucratic targets and indicators and silo-ed, goal-specific policies. These miss the opportunities to build synergies and address tensions between different goals – recognising for instance that climate action cannot be successful without addressing the unsustainability of food, land and water systems. More fundamentally, bureaucratic, silo-ed approaches miss the key insight that tackling inequality (goal 10), and gender inequality (goal 5), are cross-cutting, central to achieving all other goals. It is time to recapture the SDGs as an integrated package within a transformative agenda towards 'equitable sustainability', or 'the future we want'.

It is also increasingly clear that such transformations are Political – the missing sixth P word of the global goals? Power and politics help to uphold unsustainable development pathways – whether in the ways powerful corporate interests drive continued fossil fuel exploitation or deforestation for commercial agriculture, or the dependence of major national political regimes on polluting industries. Challenging such powerful, 'locked in' pathways, and opening up the space for alternatives, also involves an intense politics in which citizens, activists, progressive businesses, and alliances between them can play critical roles. Transformations can be led 'from above', but can also emerge from the bottom-up, seeded by people's everyday ways of living sustainably with non-human natures, practices and innovations. And to continue the pathways metaphor, small footpaths can coalesce into bigger routeways that encroach on and might ultimately displace the motorways of environmentally damaging political economies. Environment and development thinking needs to become more fully political and engaged if it is to meet the challenges of our time.

It is also increasingly clear that such politics must embrace not just the material dimensions of power – its flows of money and interests – but also the politics of knowledge with which these are often deeply interconnected. Vital questions arise about whose knowledge counts in defining 'the future we want' (and indeed who the 'we' is), and in shaping the pathways to get there. While formal scientific knowledges have key roles to play, these are never politically neutral, while recognising and nurturing everyday, 'indigenous' knowledges, including of people living in marginalised situations, can be key in fostering 'caring' transformations. In this view, transformation is not a neat, controlled process, but a more messy, emergent one, involving deliberation amongst plural pathways towards diverse, and sometimes contested, visions of sustainability and equity.

Gender has long provided a valuable lens for interrogating relationships between environment and equity, and for problematising the relationship between more bureaucratic/managerial, and more political/transformative, approaches to development. These twin struggles – or troubling intersections and contestations – have played out in diverse and changing ways through successive eras of gender-environment research and policy. A brief sketch of these would include colonial economic and environmental policies, their exploitation of both people and non-human natures, and gendered contestations and resistances to them, and then the environmental movements of the 1960s and 1970s, in which feminist and ecofeminist perspectives variously played influential roles and struggled for voice. This sketch would go on to describe the rise of 'Women, Environment and Development' (WED) approaches from the 1980s, which sought to include women (often seen as a homogeneous group) in dominant, neo-liberal development approaches which were often problematically unsustainable in themselves, and too often harnessed women's work to compensate through environmental care. 'Gender, Environment and Development' (GED) policy approaches, mirroring Gender and Development (GAD) followed, and with them more attention to gender relations in labour and resource access. The associated gender mainstreaming approaches promised a valuable integration of a gender relational perspective across all development, environment included – yet in practice was open to bureaucratisation and a tick-box approach.

Feminist political ecology from the later 1990s valuably re-emphasised the inherently political nature of gender, environment and development, including the knowledge informing it – interweaving productively with feminist critiques of science. In its more recent incarnations, 'new' feminist political ecology has fully embraced intersectionality – the ways that the dynamics of difference and (under) privilege related to gender, ethnicity, place and other factors interact in dynamically reinforcing ways – as well as acknowledging the importance of embodied, experiential forms of knowledge. Such approaches lend themselves well to informing political, transformative action towards sustainability and social justice, often built on feminist experiences and activism. Yet these politicised approaches continue to co-exist and struggle with more instrumental development policy and programme approaches. Too often, these either stereotype 'women' as environmental victims (to be helped) or 'sustainability saviours' (whose labour can be harnessed to save the planet) – or they again reduce 'gender' to a mainstreamed box in a log frame.

In this contested field, it is not surprising to find 'gender experts' in a state of trouble – facing dilemmas of identity, role, capacities for influence, and indeed dilemmas over what directions influence and change should take, for whom and why. As this book shows so vividly, and through its feminist methodology that privileges people's own voices, experiences and relationships, 'experts' deal with these struggles in numerous ways. The ongoing tensions many face between managerialism and activism, between being a political feminist and a femocrat, comes through loudly, and speaks not just to gender, environment and development but to the politics of sustainable development more broadly.

Yet, these diverse struggles and active engagement with dilemmas are also worthy of celebration. They reveal 'gender expertise' as diverse, fluid and adaptive, constantly crossing boundaries and providing bridges between research and policy, knowledge and action, policy and politics. They highlight how often the best and most useful expertise is also the most negotiated, reflexive, and humble. These are hugely valuable qualities in people and professionals involved with environment and development, and especially so in an era where sustainability needs plural, politicised transformations. Gender expertise that is self-aware about its troubles could actually be part of the politics of hope that the world now needs. So rather than wish them away, let's keep the dilemmas and troubles brought to light so vividly by the contributors to this book alive, and treat them as a source of inspiration in charting pathways that work for sustainability and social justice.

Professor Melissa Leach CBE FBA MAE
Director, Institute of Development Studies
Member, Science Advisory Council (SAC),
Stockholm Environment Institute (SEI)

Preface

This is an exciting, rich and engaging book on the troubled engagement of Feminist Political Ecology (FPE) with the professional world of technical environment and development organisations. It captures how 'gender experts' working on gender, environment and development are interpreting and translating global narratives on sustainable development in their daily work. The book candidly depicts the editors' and authors' struggles and achievements, as well as deep seated resistance and ambivalence to the term 'gender expert'. It does so by building on decades of feminist critique of science and technology, knowledge production and shared reflections on how feminists become, often reluctantly, implicated in colonial global north–south relationships, racialised positions of power and authority that intersect with the epistemic hierarchies around science and the bureaucratic status quo.

As the brilliant introduction by the editors Bernadette Resurrección and Rebecca Elmhirst indicates, these voices from FPE come from a depth of analysis and understanding of the tensions and ruptures in the expert gender discourse in the world of technical policy on sustainable development and environment. The authors draw on methods and approaches from science and technology studies, feminist international relations, feminist studies and critical development scholars, to show how they engage with the messy, fragmented and everyday subversions of feminists (gender experts) working in the technical environment and development world. The book shows how the critical discourse about the gender expert in development has moved beyond the issue of co-optation or the do's and don'ts of gender mainstreaming to examine the subjectivities, positionalities and relationships embodied by gender experts as they work through the everyday dilemmas of doing gender work in technical environment and development contexts.

The book's theoretical framing positions it at the forefront of embodied theorising in FPE. The key question – 'from who and where are we, rooted in place, networked across space and time, and connected to resources' – is answered in personal accounts, centred on the feminist practice of reflexivity that allows the contributors to reflect on power in the construction of their work and selves as 'gender experts'. The understanding of 'gender expert' as it is explored in the book, personally and in dialogue with the editors, shows how gender expertise is produced in complex ways, as the editors state 'through power-laden webs of political and social relationships that define and create expertise in gendered, racialised and colonialist ways'.

The editors in their expansive introduction, point to three very useful ways of understanding the subjectivity of the expert and the boundaries of expertise and the potential for meaningful (feminist) change. They look at how gender expertise is produced with and in contrast to feminist (social movement) politics; the critique of western expertise by postcolonial (and decolonial) politics; and in the context of competing truths of scientific epistemologies. This reading of gender expertise enables the authors to speak to productive tensions as they negotiate their position as gender experts that have historical meaning in terms of mainstreaming gender in environment and development policy, and at the same time it recognises the struggles for feminists to claim and indeed to refuse what it means to be a 'gender expert'.

What is particularly insightful in this approach is the attention paid to the changes over time as the authors describe how they have embodied the position of 'gender expert' moving between feminist and bureaucratic spaces in social movements, university, development organisation, or advocacy organisations. Understanding these trajectories through the stories told by the authors shows the fluidity of 'gender expertise' and how credibility can be required in more than one space, along with the ability to 'converse' in the language and practices of different institutions, networks and disciplines.

Another insight is how these negotiations are not only in terms of the embodied experience of the authors but also in discourse. While approved modes of gender, environment and development-speak – positivist scientific surveys or catchy messages with gender statistics, simple compelling narratives – might be strategic in order to cross the different technical and social policy spaces, it is vital to go deeper. This going deeper requires attention to the everyday, embodied and emotional relations of women, men and communities and their knowledge and practice of gender, environment and development. The book shows the small successes which come from recognising local, historically embedded and culturally grounded forms of knowledge. At the same time, it underlines the importance of the larger battle for epistemic justice and to engage with powerful environmental institutions by understanding systemic injustice from an intersectional perspective.

The book is also highly original, and daring, given the institutions in which most of the authors are working, by generating knowledge collaboratively through a process of reflexive conversation and embodied theorising. By writing and thinking from the embodied experience, the authors present in their self-reflection and collectively question their situatedness within different axes of power, privilege and affective relations. By adopting this feminist methodology, the book opens up new ways of understanding debates around gender, environment and development, women's empowerment and gender mainstreaming. The authors carefully position themselves in the shaping of gender, environment and development discourses in order to bring out how gender expertise is built in relationship with feminists in social movements; in the collective questioning of colonial legacies and racialised, class and other forms of privilege and authority; and in negotiating what counts as admissible knowledge according to which discipline and ways of knowing.

By 'staying with the trouble' the book allows us to understand how productive tensions are necessary if we work with the plurality of feminist knowledges and practices among activists, academics and development practitioners in the business of, and resistance to, the doing of gender, environment and development. The book's troubling of the everyday, the institutional and the political, breaks down the authority of gender expertise that is assumed to be based on universal gender knowledge showing how these authors are navigating and changing the politics of coloniality and patterns of power which underlie gender, environment and development discourse.

The book is a fascinating collaborative autoethnography in FPE that emerges from what the editors describe as 'polyvocal essays' that build on the 'co-learning encounters' and 'embodied listening' in the process of writing the book. The conversations traverse science and environmental discourse and institutional practices as well as deeper reflections on epistemologies and power. The book goes far beyond a dull, plodding survey of what gender expertise means in environment and development. Instead it is a lively, engaging set of conversations that unflinchingly point to the tensions and difficulties of this slow revolution to bring gender to environment and development science and policy, and beyond.

Professor Wendy Harcourt
International Institute of Social Studies,
Erasmus University Rotterdam
The Hague

Acknowledgements

I owe deep thanks to the Stockholm Environment Institute (SEI), which encouraged me and gave me the professional space and support to co-write this book. I was inspired by SEI's core mandate of 'bridging science and policy', which prompted me to seek what this actually means in real-world, human and political terms. I am grateful to the book's contributors traversing the environment and development world, who so generously and honestly shared their reflections on their uneasy but skillful engagements: championing equity, care ethics and justice in spaces where these are not easily accommodated or amenable to neat forms of predicting, planning, evaluating and control. Without them, this book would have just remained an aspirational idea. Thank you to my esteemed colleagues in the Gender, Environment and Development Research Cluster at the SEI Asia Centre with whom I've journeyed the byways and narrow streets of translating feminism at SEI and beyond, and our Centre Director, Niall O'Connor, who patiently supported my need for headspace for this book in the midst of the daily and busy stream of work challenges. Big thanks to my Gender & Social Equity Programme colleagues Natalia Biskupska and Rajesh Daniel who helped organise the contributor conversations and later made useful edits and suggestions to drafts in progress. Special and deep gratitude to Jonathan Shaw, who had to take a few holiday breaks alone because I had to 'put life on hold' so that the writing could inch forward. This book is especially dedicated to Ayra Jean Sayo Leung, who might one day think about her strange grandmother who feverishly multi-tasked as she awaited her (long delayed) birth whilst starting a book on feminist political ecology in the waiting lounge of a Hong Kong hospital. May her world be filled with feminist caring and kindness that ours today so badly needs. Finally, I am profoundly grateful to Becky – Professor Rebecca Elmhirst – comrade-in-arms for many years with whom I spent many delightful hours discussing book angles and ideas, re-working text together, as we navigated this unbeaten thematic track for us. In her, I saw and learned about unwavering commitment, empathy and intellectual fortitude. Becky truly embodies the best qualities of feminism in the academy.

Bernadette P. Resurrección, Bangkok

I am very grateful to the University of Brighton for the encouragement and support I have had in working on this book. Present and former colleagues deserving a special mention include Jenny Elliott, Kath Browne and Kirsty Smallbone – warriors in their own ways. I am lucky to be situated in a school that brings together researchers and teachers from wildly different disciplinary starting points, and this has challenged me to find ways of working meaningfully across science and feminist political ecology, alongside supportive colleagues in Brighton's Centre for Spatial, Environmental and Cultural Politics. I am grateful for the opportunity to learn from all the contributors to this book, whose daily achievements at the sharp end of environment and development practice are inspirational. I thank them for generously answering my clunky academic questions, and the fruitful discussions these provoked. I have been enormously fortunate to be part of WEGO-ITN (Wellbeing Ecology Gender and cOmmunity), an EU Innovative Training Network which received funding from the European Union's Horizon 2020 research and innovation program under the Marie Sklodowska-Curie grant agreement No. 764908-WEGO 2018-2021.WEGO has opened up extraordinary new pathways for 'doing feminist political ecology' alongside inspiring feminist political ecologists from around the world. From this, special thanks to doctoral researchers Elona Hoover, Dian Ekowati, Alice Owen and Siti Maimunah and the Extracting Us project. Big thanks are due to my family: Gordon has always supported my work, but he will be grateful to have his kitchen table back. I dedicate this book to him and to my children Angus and Shona with all my love and care as they embark on their adult journeys. Finally, my biggest thanks go to Babette – Dr Bernadette P. Resurrección – who turned her reflections into an idea for a book, and whose scholarship, tenacity, friendship and care nurtured the project to a conclusion, tending the young shoots of the manuscript when I had to step out for elder care. She has been my inspirational fellow traveler for more than two decades. From Babette I have learned what it means to cultivate feminist political ecology in unyielding ground, and how to wield the hoe most effectively. May we continue to learn together.

Rebecca Elmhirst, Brighton

Together, we are most grateful to the Routledge team behind the *Gender and Environments Series,* Matthew Shobbrook, Annabelle Harris, Kristina Wischenkamper, and Susan Buckingham, for their competence and patience, steering us through the completion of this book.

The views expressed in this publication are the personal views of the editors and contributing authors and do not necessarily reflect the views or policies of the organisations with which they work.

Introduction: Troubling gender expertise in environment and development

Voices from feminist political ecology

Bernadette P. Resurrección and Rebecca Elmhirst

1. Why this book, why now?

This book is about the daily struggles and achievements of 'gender experts' in the environment and development world. With a geographical focus on the Global South, we explore their work in environment and development organisations and interventions, where they are charged with advancing gender equality and social equity and aligning this with visions of sustainable development. Global climate change and disaster agreements and protocols contain clauses for the integration of gender equality in intervention programmes, building on multi-lateral conventions such as the United Nations Sustainable Development Goals (SDGs) and the 2015 Paris Climate Agreement, and earlier platforms such as the United Nations Commission on Environment and Development Agenda 21, the Convention on Biological Diversity and Section K of the Beijing Platform for Action. In the post-2015 global development agenda, 'gender' has emerged as an important issue in global environmental governance and in the programming of environment and development organisations, driven by a requirement to address the 'gender gap' and its social and economic consequences (Arora-Jonsson & Sijapati Basnett, 2017). In this framing, addressing gender inequality is seen as necessary for managing environments and reducing poverty effectively and efficiently (Chant & Sweetman, 2012; Prügl, 2015). Access to donor resources for environment and development work is often conditional on programmes explicitly demonstrating adherence to 'gender mainstreaming' or 'gender-inclusivity' principles and actions. The scope of this work is vast, ranging from gender analysis in the planning of technical projects on climate change, disaster risk reduction, energy transitions, pollution and food, crop production, land and water insecurity, to the integration of gender variables in geospatial technologies, modelling, scenario- and futures analyses and climate services.

The 'doing' of gender mainstreaming or gender-inclusive work has led to the emergence of the 'gender expert', a figure that is central to the professionalisation of 'gender' within the realm of environment and development organisations, programmes and interventions. We place the term 'gender expert' in parenthesis here as we deliberate on its emergence within the knowledge politics of environment and development work. Broadly, the job of the gender expert or gender advisor includes the review of policies and strategies, training, skills development and awareness-raising and implementing planning and monitoring systems to support gender equality (Eyben & Turquet, 2013; Davids et al., 2014; Bustelo et al., 2016; Caglar et al., 2013; Kunz & Prügl, 2019). The rapid rise of gender expertise in recent decades has come about in response to the wide adoption of gender mainstreaming in international organisations and governments. An international cadre of gender experts has emerged, comprising feminists, consultants and trainers who play a key role in translating feminist knowledge into development policy and practice (Prügl, 2013).

The rise of gender experts and gender expertise in development has prompted a vibrant critical discussion and raised some disquiet that feminism has lost its radical edge, as feminist agendas are co-opted for government purposes and even corporate agendas (de Jong & Kimm, 2017; Prügl, 2015; Cornwall, 2007a). A self-reflexive literature by gender experts themselves highlights the depoliticising effects of gender mainstreaming and failures to achieve its envisaged goals (Miller & Razavi, 1998; Baden & Goetz, 1997; Cornwall, 2007b; Eyben, 2010; Davids et al., 2014; Parpart, 2014; Milward et al., 2015). Contributors to this debate cite an overemphasis on technocratic approaches that rest on measurability, quotas and simplifications that are embedded in the development logic of international organisations, which means the transformative intent of gender mainstreaming is 'streamed away' (Mukhopadhyay, 2004: 95; Kothari, 2005). Similarly, its framing through a neoliberal discourse with a vocabulary of effectiveness, efficiency, impact assessment and 'smart' economics, diminishes any power to address feminist agendas (Cullen & Murphy, 2018; Ferguson, 2015; Nirmal, 2016). Instead, gender mainstreaming 'adds value' to 'existing meta-narratives' (Mama, 2004: 123), with a danger that underlying structural inequalities and wider well-being concerns are obscured (Fraser, 2009; Bradshaw, 2013; Chant & Sweetman, 2012; Calkin, 2017; Cornwall & Rivas, 2015; de Jong, 2016).

In this book, we explore the possible tensions and ruptures that are involved in bringing the feminist knowledges that underpin gender mainstreaming into the realm of technical and scientific knowledges that inform the work of environment and technical development organisations. Within the latter, technical, evidence-based and managerialist approaches dominate, based on the use of statistical indicators and impacts, hard evidence and rigorous data, weights in costs and benefits, imaging and remote sensing technologies, and risk assessments to render (unruly) nature more intelligible and eventually, more amenable to control and management (Taylor, 2014; Eyben, 2013; Valdivia, 2015). We join Arora-Jonsson and Sijapati Basnett (2017) in being struck by the enthusiasm with which 'international environmental organizations appear to be increasingly adopting "gender mainstreaming" and employing "gender experts"'(p. 309), despite the reverberations of the critiques outlined above. In part, this reinvigoration is being driven by the requirements of major multilateral donors with a mandate to address Sustainable Development Goal 5 (gender equality), and from the influence of bilateral donors such as Sweden that have been prominent in providing funds and technical assistance in environment and technical spheres.

We ask whether the epistemological interface between gender mainstreaming knowledges and environmental science raises particular questions around the subjectivity of the expert and the boundaries of expertise and the potential for meaningful (feminist) change. Critically, for the 'gender expert' positioned within environment and development organisations and programmes, the loss of a 'transformative edge' may be even more pronounced as gender work involves them operating in techno-scientific and managerialist fields with epistemologies and logics that amplify the instrumentalisation, simplification and depoliticisation highlighted in gender mainstreaming critiques. We debate this with an awareness that the entry of 'gender experts' into technical environmental fields has been hard-won. It represents the creation of a space in which at least there is dialogue for gender equality and scope for the making of a slow revolution to social justice in the realm of environmental work (Davids et al., 2014: 397; Arora-Jonsson & Sijapati Basnett, 2017).

Analyses of gender mainstreaming that unpick performative framings of power and subjectivity (e.g., Davids et al., 2014; see also Butler, 1990; Nightingale, 2011) and those that elicit personal and self-reflexive accounts of 'doing' gender in development contexts (e.g. Eyben & Turquet, 2013) have each challenged the supposed dichotomy between 'good feminism' (and 'authentic feminists' with transformatory intent) and 'co-opted feminism' (of 'femocrats' charged with gender mainstreaming). In different ways, these projects have punctured any sense of a simple binary by pointing to forms of 'slow revolution' that can ensue. Change comes through the 'small, messy, fragmented and everyday kinds of subversions, conscious and unconscious' (Davids et al., 2014: 404) where gender professionals are 'subversively accommodating' (Eyben, 2010), and the radicalness of change more deliberately relates to the depth of engagement (van Eerdewijk & Davids, 2013). Eyben and Turquet's (2013) edited collection not only gives voice to these instances of 'power in recoil' (Nightingale, 2011; Butler, 1990) but its contributing authors provide reflexive accounts of their business as gender experts navigating the professional cultures within development organisations. What opportunities for power in recoil might there be in more technical settings?

Our interest lies in exploring the specific challenges that gender experts face in embarking on this slow revolution in contexts dominated by applied environmental science. Does this raise a particular set of questions and concerns that deserve scrutiny and self-reflection? Might we query the role of the gender expert in this context: transforming what, for whom and with what authority? Partly in response to critical academic debates around gender, environment and development theory and practice (Leach, 2015; Resurrección & Elmhirst, 2008), environment and development organisations of various kinds are undertaking efforts to stock-take and reflect on whether and how gender discourse and feminist theory have led to positive effects in areas within environment and natural resource management research and practice, and beyond that, in addressing sustainable development goals (Resurrección et al., 2019; Cochrane & Rao, 2019). Our aim is to extend and complement this work and to respond to the challenge that at this juncture we need 'an inquiry into the internal dynamics of (environment and) development's "regimes of truth" or of the production of professional identities' (Mosse, 2011: 2; parenthesis ours); and 'more research on knowledge producers, practitioners and policy makers – to understand how we work and our own preconceptions' (Arora-Jonsson, 2014: 11). We do this by engaging the insights drawn from a rich literature on gender experts and other professionals in development contexts (Prügl & Lustgarten, 2006; Eyben & Turquet, 2013; Davids et al., 2014; Kunz et al., 2019) and bringing these into conversation with perspectives drawn from the conceptual terrain of feminist political ecology (hereafter FPE), with its emphasis on embodied power and the politics of situated knowledge (Haraway, 1988; Rocheleau et al., 1996; Resurrección & Elmhirst, 2008; Sundberg, 2017; Ahlborg & Nightingale, 2017). FPE provides the framing tools for our focus on gender expertise and the tensions arising in the co-production of knowledge as gender experts operate in environmental organisations that are charting futures in food security, climate change adaptation and mitigation, sustainable energy, biodiversity, disaster management, water security, low-carbon transitions, environmental governance, and coastal management under neoliberal conditions.

As we ponder on gender expertise and mainstreaming and their interconnections with feminist political ecology in technical environment and development realms, we also engage with feminist science, technology and society perspectives (STS). Feminist STS studies have identified processes by which truth claims made by scientists enact an 'objective' perspective that is constructed as inherently valuable (Harding, 1986; Haraway, 1988; Bauchspies & Bellacasa, 2009). This raises some important questions. Is it possible to remain committed to a transformative agenda, whilst navigating environment and development bureaucracies *and working across* an epistemological interface with positivist science? Does an applied science professional context add an additional layer of complexity for gender experts whose 'object' of expertise (gendered power relations) is assumed to lie outside the purview of natural and technical science (Harding, 1986)? What kinds of conflicted subjectivities, positionalities and relationships are embodied by gender experts as they work through everyday dilemmas inherent in doing gender work in these technical contexts?

We thus explore the tensions that gender experts face as they seek transformative social change whilst working in technical-environmental contexts that demand simplifications and technical fixes (Li, 2007; Mosse, 2011). As recent global agreements attempt to accelerate action on climate change and bring renewed urgency to tackling environmental challenges, gender experts – advisors, researchers, organisational focal points – are called upon to navigate the complex epistemological ruptures and coalitions that emerge when gender analysis is brought into technical, environmental management settings fortified by positivist science. We are interested in understanding how people come to be 'gender experts' in such contexts, the authority and epistemic agency they are able to engage within environment and development work, and the extent to which they can carve out opportunities for 'slow revolution' towards social transformation, environmental justice and gender equality whilst working collaboratively with scientists to address profound and existential environmental threats. By reflecting on and unpacking the experiences of gender experts and their everyday professional lives, our discussions also re-visit the suite of technical planning tools and other cultural artefacts of gender expertise (Liebrand, 2014) that have been developed within the 'gender business'. This provides an opportunity to reflect on whether they have led to meaningful gender equality outcomes or have blunted the possibility of deeper transformation.

The book presents a collection of polyvocal essays written together with 'gender experts' working in technical and environmental science-led organisations, whose work is for the most part geographically located in the global South. We see this as an example of embodied theorising in feminist political ecology, reflexively examining our work in gender and environment 'from who and where we are, rooted in place, networked across space and time, and connected to resources'(Rocheleau, 2011; Harcourt & Nelson, 2015). Through our discussions and collaborative reflexivity, the chapters offer a series of reflections on our distinctive, situated and embodied experiences and shifting, troubled identities as 'gender experts' in the field of environment and development. As the chapters show, not all of us regard ourselves as 'gender experts' and some of us refuse this label entirely. What is shared is a willingness to deliberate on uneasy engagements with positivist epistemic communities and to begin to unpack the politics of accommodating gender in science and environmental discourse and institutional practices, and through this, touch on promising pathways towards a more politicised and emancipatory feminist praxis in environment and development. Below, we set out the approach taken in this book, beginning with a discussion of how we are engaging ideas associated with feminist political ecology and feminist STS studies. This provides the conceptual moorings that guide our reflections as well as a framework for our reflexive authorial approach.

2. Framing through feminist political ecology

Our discussions are anchored within feminist political ecology (FPE): an expansive and open-ended field rooted simultaneously in political economy, ecology and feminist activism (Sundberg, 2017; Elmhirst, 2015; Harcourt & Nelson, 2015). FPE research and practice broadly seeks to illuminate and act on social and gender relations of power associated with natures, cultures and economies, and in doing so, questions market-driven neoliberal forms of development. FPE challenges dominant practices of knowledge and authority that cast environmental change as a politically neutral process amenable to technical management (Rocheleau, 2015; Resurrección, 2017). Thus, FPE has emerged from earlier gender, environment and development studies (Elmhirst, 2015), and been enriched and given a more transformative flavour through strands that originate in environmental justice and feminist social movements (Di Chiro, 2016). Moreover, FPE is informed by feminist science, technology and society approaches that emphasise power and the situatedness of knowledge claims (Haraway, 1988; Rocheleau, 2011; Nightingale, 2011).

Our reasons for engaging with FPE are many. First, FPE is an umbrella for the reflexive, conceptual and political tools that can push back against the charge that gender equality efforts are reduced to being a bureaucratic obligation. Feminist political ecologists are also among those who have been dismayed by the apparent professionalisation of the feminist movement and its apparent co-optation within both public and private sector development bureaucracies (Calkin, 2017; Lanz et al., 2019). Whilst one response is to distance FPE from mainstream development practice, with its link to neoliberal modes of governance and institutionalism (Mosse, 2011), another response is to 'stay with the trouble' (Haraway, 2016) in order to recognise and hold in productive tension the plurality of feminist knowledges and practice among activists, academics and development practitioners (Bustelo et al., 2016; Holvikivi, 2019). We follow others in troubling the boundaries between these different subject positions: we see gender experts as 'activists' in the organisations in which they work in large part because they broadly advance agendas of social inclusion. Furthermore, as Kunz et al. (2019) note, the situatedness of gender experts means some are deeply immersed at a personal level in work that challenges the coloniality of universalised gender knowledge in development contexts (Kunz et al., 2019). Thus rather than characterise separate worlds of feminist activists on the one hand and gender experts and femocrats on the other, the aim must be to reflexively interrogate the silences between epistemologies and ontologies amidst this plurality (Nightingale et al., 2019).

Secondly, the transformative socio-ecological justice aspirations of FPE may be served by continuing to engage with mainstream environment and development practice, particularly as expressed through the business of international environmental organisations with their institutional capacity for making change happen. We join others in arguing that there is a real need for FPE to engage with powerful environmental institutions, particularly where the adoption of gender mainstreaming 'is clearly having political effects beyond academic and feminist communities' (Arora-Jonsson & Sijapati Basnett, 2017: 311). A number of those occupying the position of 'gender expert' in technical organisations regard themselves as feminist political ecologists, steeped in this subfield's knowledge networks. This engagement is therefore already happening at an intimate and embodied scale, as some of our contributors demonstrate.

Thirdly, FPE offers critical tools that simultaneously serve the aims and objectives of the book, frame our discussions with our contributors, and inform our approach and process. We examine questions of subjectivity and power in the making of 'gender experts' (Butler, 2004; Sundberg, 2017), before drawing an analysis of the ways in which gender has been 'rendered technical' (Li, 2007) in environment and development contexts, setting the parameters for much of the 'gender work' of our contributors. Our concern is to explore the epistemological boundary between gender theories and the varied but largely positivist science frameworks guiding the work of technical and environmental organisations. This draws us to conceptualisations of scientific authority, located in FPE's engagement with feminist science, technology and society debates (Haraway 1988; Harding, 1986; Nightingale, 2011).

Finally, the methodology and approach to authorship taken in this book is informed by feminist approaches to collaborative writing (Nagar, 2014; Harcourt, 2016), which has found resonance within FPE. This involves the co-production of often highly personal accounts, centred on the feminist practice of reflexivity, understood here as a form of disciplined self-reflection and awareness of the context in which knowledge is produced and situated (Haraway, 1988; Lumsden et al., 2019). This form of knowledge and writing is seen as lying outside mainstream science and political ecology (Armiero et al., 2019), even though it has been a longstanding mode of analysis and expression in academic feminism. We see reflexivity as important in highlighting key dimensions of gender expertise and the construction of knowledge across epistemological boundaries, and in recognising one's proximity to power (Bustelo et al., 2016; Ferguson, 2019). Further, we echo Bacchi (2009) in seeing reflexivity as a potential strategy to capitalise on small margins for change. Cognisant of the volume of criticism that has lodged the loss of feminism's political edge with gender mainstreaming and the professionalisation of gender knowledge, this approach recognises the value of the grounded practices of gender experts in environment and development settings (Cornwall et al., 2007; Verloo et al., 2009). Together, we explore the uneasy decisions and positionalities that each of our contributors has to create in order to engage in the various spheres of political influence required in order for gender equality agendas to gain serious traction (Ferguson, 2019).

3. Making the 'gender expert' and gender expertise

Our starting point has been to instigate conversations with people working in a professional capacity as 'gender experts' within technical environmental organisations that ranged from research institutes, advocacy-led non-governmental organisations, bilateral donors and policy think tanks. In identifying our contributors, we were effectively beginning with a two-tiered definition of the 'gender expert', as (i) a person with 'knowledge regarding the cause-and-effect relationship between policies, actions and/or activities and gender inequalities', who (ii) 'is formally requested to provide her (sic) knowledge and service' (Hoard, 2015: 12). However, as each of our contributors was, to varying degrees, uncomfortable with the label, we ask what and who is a 'gender expert'. In defining 'experts', sociological accounts of professions focus on the ways experts are defined by their acquisition and possession of specialist knowledge. They seek to secure their monopoly over knowledge and its use through credentialist ideologies, such as formal qualifications, and socialisation into the terminologies and social rules of the profession, and in doing so, distinguish 'expert' from 'everyday' gender knowledge (Saks, 2012).

However, Bustelo et al. (2016) state there is no agreed definition of the term 'gender expert', no common criteria of what their work entails, and no systematic mapping of the profession from which its boundaries might be defined, and from whence claims to authority may come. Instead, the lines defining what and who is a 'gender expert' are blurred and gender expertise is only weakly professionalised

(Prügl, 2013). Even as the institutional requirement to establish 'gender mainstreaming' is based on a set of practices that require specialist knowledge and training enshrined in specific claims for authority, there are no explicit standards, training or expectation of prior knowledge for gender experts, and only a minority has a degree in gender studies (Bustelo et al., 2016; Prügl, 2013; van Eerdewijk & Davids, 2013). Gender experts together may form what might be described as an epistemic community based on recognised expertise, competencies and authoritative claims (Hoard 2015) but their routes into this are varied, either via specific training on gender or via other kinds of social fields that led them to become gender advisors.

In line with a sociological emphasis on professionalisation, specialist knowledge and authority, a literature has emerged from discussions of gender mainstreaming that focuses on the everyday trials and tribulations of 'gender experts' in gaining authority within the organisations in which they work, and in working across epistemological boundaries, with the quantitative social sciences (i.e. economics) and the natural sciences (Eyben & Turquet, 2013; Ferguson, 2015). Feminist sociologies of the workplace have also anticipated the observation that knowledge hierarchies within organisations place a relatively low value on gender expertise – gender professionals may be early career employees with limited authority within organisational hierarchies, or women whose 'main' workload is topped up by gender mainstreaming responsibilities within their workplace. This, along with other factors, can mean gender experts may have limited authority within their organisations, whilst gender expertise itself is awarded low economic value (Ferguson, 2019). For example, in their discussion of gender mainstreaming in an international environmental organisation, Arora-Jonsson and Sijapati Basnett describe how despite the gender leader having a place on the organisation's management board, the gender programme was 'tested' by outside consultants, unlike the other programmes (Arora-Jonsson & Sijapati Basnett, 2017).

Recent debate has focused on the *contested* positions of those whose work involves the translation of feminist and gender theories to practitioners (Bustelo et al., 2016; Davids et al., 2014). 'Gender experts' are positioned either as 'trojan horses' for feminism within development organisations or as 'femocrats', complicit in bureaucratising feminism and in the loss of its transformative edge (Cornwall et al., 2004). Others have troubled this apparent boundary, showing how people working as 'gender experts' in fact occupy multiple subject positions in and through their daily practice (Kunz & Prügl 2019; Ferguson, 2015; Eyben & Turquet 2013). In other words, complex subjectivities are materialised in the context of gender mainstreaming work.

Theorisations of subjectivity and power that figure in Feminist Political Ecology (Butler, 1990; Nightingale, 2011; Bee, 2019) direct us towards examining how expert subjectivities and expert knowledges (of gender) are performative and co-produced. In this view, subjectivities (such as the 'gender expert') are not stable but come into view and are reproduced relationally, through power-laden webs of political and social relationships that define and create expertise in gendered, racialised and colonialist ways. The embodied subjectivity of the gender expert is thus lived and enacted through these webs of relationships in material and institutional spaces (environmental organisations, workshops, laboratories, networks, in fieldwork contexts) and through everyday practices (conversations, training, advice and advocacy) materialised in the tools and artefacts of gender mainstreaming.

A useful framework to consider the performativity of expert subjectivities and expertise is developed by Kunz et al. (2019), who focus on the boundaries that gender experts must navigate, draw and erase as they seek to get heard and establish their authority. Kunz et al. (2019) identify three 'fronts of struggle' that play a role in forming the subjectivity of gender experts and in the forging of gender expertise in development settings: first, the boundary between gender expertise and feminist (social movement) politics; secondly, the relationship between gender expertise and postcolonial (and decolonial) politics; and thirdly, struggles around the authority of gender expertise in the context of scientific epistemologies.

Taking the first of these, the relationship between gender expertise and feminist politics is multi-stranded and multi-directional, and this complexity underpins the conflicted subjectivity of gender experts. In interviews undertaken for their study of gender professionals, Kunz et al. (2019) find an ambivalence towards political or social movement feminism, as people either did not self-identify as feminist or drew a distinction between their personal feminist politics and professional gender expertise. This kind of boundary making is understood as a way of keeping their gender work safe from feminism (Kunz et al. 2019), seeing politics as antithetical or unhelpful towards the 'do-ability' of their professional responsibilities. The ambiguity lies in the acknowledgement that the knowledge on which gender expertise rests is derived from political feminism. Earlier debates have noted the ways in which stereotypical understandings of feminism in many institutional contexts make this a pragmatic decision (Cornwall et al., 2007) but it is one that reinforces a distinction between feminist social movements and the figure of the 'femocrat' or bureaucratic 'mandarin' that have been subject to much critique (Miller & Razavi, 1998; Cornwall et al., 2007).

Boundary work is embodied in specific ways, which challenges a clear-cut dualism between bureaucracy 'insiders' and feminist movement 'outsiders' (Rao, 2006). Discursive power within development organisations and in feminist social movement politics work together to produce gender expert subjectivities that are placed as 'outsiders' within their own organisations because their expertise is at odds with dominant expert paradigms, whilst feminist politics places them as bureaucracy 'insiders' or femocrats (Eyben & Turquet, 2013). Eyben and Turquet adopt the phrase 'tempered radicals' to describe bureaucratic subjectivities that draw on support from friends and colleagues in the feminist movement. Thus, boundaries not only bound difference (Nightingale, 2011) but include relations of proximity, positional slippages and political solidarities of various kinds (Holkiviki, 2019; Harcourt, 2016). A performative approach to expert subjectivities gives space for showing the instability of boundaries and embodied subjectivities (Butler, 2004; Nightingale, 2011). Temporal and spatial dimensions of power mean bodies can potentially move between feminist and bureaucratic subject positions over time (over the life course) and across spaces (working in university, development organisation, or advocacy body), thus troubling the boundaries of 'gender expertise' and its separation from feminist knowledge.

The second area of boundary work relates to understanding how the subjectivity of gender experts and the authority of gender expertise is shaped through the embeddedness of gender and development expertise in the politics of coloniality, defined as longstanding patterns of power which redefine knowledge production 'in ways that accredit the superiority of the coloniser' (Mendoza, 2015: 114; Mohanty, 1988). Such patterns of power are related to the establishment of the universal authority and hegemony of Western knowledge and discourses (Sanna & Varikas, 2011: 7, cited in Verschuur & Destremau, 2012). The figure of the gender expert is implicated in the politics of coloniality as the professionalisation of gender expertise in international development institutions has been built on a universalised gender knowledge, assumed to be transferable and applicable in any given context (Kunz et al., 2019). As Kunz et al. (2019: 34) note, 'a geographical hierarchisation of gender knowledge privileges supposedly universal insights and problem-solving skills over socioculturally specific gender knowledge and solutions emerging from a particular context', whilst the authority to 'identify relevant and complex gender issues in the hands of gender experts is located in big international institutions'. Yet the ways in which this plays out in and through expert subjectivities is complex. By embodying intersectional articulations of racialised, classed and gendered power, gender experts are variously positioned in relation to the logics of coloniality in gender and development within their personal and professional biographies, complicating their authority as 'experts' in their everyday practice. The logics of coloniality draw a distinction between 'international' and 'local' staff, and the everyday relational dynamics between these two positions defines the 'gender expert' and their field of expertise in different institutional contexts.

Thirdly, practices that mark and attempt to dissolve the boundary between gender expertise and scientific knowledge are part of the processes implicated in both establishing expert subjectivities and the field of gender expertise itself. Kunz et al. (2019) suggest that a key strategy for gender experts seeking to gain authority within development organisations has been to erase the boundary between epistemologies of gender and of science, the latter characterised by quantitative knowledge and the employment of positivist methodologies. This has not only meant utilising statistical indicators and quantitative impact data, but also subscribing to the logics of practicality rather than critical thinking. 'Practicality is knowledge that is able to provide clear-cut solutions amenable to technical manipulation and intervention' (Kunz & Prügl, 2019: 31). A dominant development vocabulary of effectiveness, efficiency, impact assessment and 'smart' economics is put into play (Davids et al., 2014: 401), as gender experts translate their work into language that can gain traction and hold weight in the contexts in which they work (Dersnah, 2019). For example, Kunz et al. note that gender experts in the World Bank need to know economics and thus to use particular kinds of data (Kunz et al., 2019).

In organisations whose operations are underpinned by environmental sciences, gender experts seek to establish their authority by translating their work through governing technologies such as log frames, surveys and gender audits, which 'reflect the episteme of the biophysical and natural sciences' (Mukhopadhyay & Prügl, 2019: 10). Complex subjectivities emerge as gender experts who are tasked to 'integrate' gender with natural science research and knowledge production attempt to challenge epistemic hierarchies that position their expertise as an appendage to the positivist sciences. Rather than reject the cognitive authority of science, gender experts (like social scientists working in multi-disciplinary environments) gain authority and credibility by developing a working fluency in the language and practices of dominant networks and disciplinary specialisms within their organisations (Forsyth, 2003; 2011). Thus gender work is made 'do-able' within an overarching Cartesian logic of making (unruly) nature intelligible and manageable through technical means (Taylor, 2014), in pursuit of a productivist market logic.

Processes of boundary making and undoing are inscribed and embodied within gender expert subjectivities in other ways also. Hiring practices for gender experts within environment and development organisations favour those candidates with an academic background or a working knowledge of a relevant academic discipline (e.g. marine biology, conservation science, forestry or engineering, depending on the context). Those with a background in social science or gender studies achieve something similar by undertaking technical training, familiarising themselves with the technical requirements or the field, positioning themselves as knowledge translators that are adept at working across hierarchical epistemological boundaries and tuning in to specific socio-environmental contexts. In so doing, their subjectivities blur epistemic distinctions and the universalising logic of gender mainstreaming.

4. Gender experts engage with 'science'

In exploring the uneasy engagements of gender experts with scientific authority in technical environmental settings, we draw on insights that emerge in the interface between FPE and feminist STS. We now see a growing number of gender experts in technical environmental organisations as current global environmental agreements are opening up to norms upholding gender and social equity as mentioned at the beginning of this Introduction. We note that the technocratic aspects of gender mainstreaming – that have been subject to so much feminist scrutiny – are coalescing with positivist frameworks that guide applied science, technology and environmental research and policy. We now bear witness to an emerging coalition of knowledges, reintroducing gender through the neoliberal economic discourse that colours the gender mainstreaming project with a dominant development vocabulary of effectiveness, efficiency, impact assessment and "smart" economics' (Davids et al., 2014: 401) that sidesteps analysis of structural inequalities (Desai, 2007). This transformation has enabled technical organisations to welcome gender expertise – and its managerialist and economistic turn – into their assemblages of modelling, GIS, and other types of quantitative indices to track, measure, evaluate and predict environmental outcomes and trends (Cornwall, et al., 2007).

The ascendancy of science in shaping environmental agendas comes with the aim to curb the increasingly catastrophic consequences of environmental and climate change largely through technical fixes (Nightingale et al., 2019). This is reflected in the growing number of science, technology, engineering and mathematics (STEM)-oriented experts in key advisory roles for governments and multilateral agencies committed to figure ways out of the environmental and climate crises. This is especially felt in the governance of science and technology for sustainable development at different scales. For instance, at the UN Environment Program (UNEP) alone there exist 17 international global environmental regimes with each of their global and regional secretariats that have established their respective scientific committees and partner think tank organisations. This has driven renewed attention to examining the efficacy of the science–policy interface (Gluckman, 2016). Despite its uneven and unpredictable receptivity and even denials from some in the policy world and other quarters of conservative scepticism, science has nevertheless been vested a distinguished place in large part 'by its relative objectivity obtained through formal processes designed to limit bias in data collection and analysis' (Gluckman, 2016: 969).

In response, STS and political ecology scholars have thus been motivated to unpack the social making of science in a number of ways beginning by questioning the almost universal adherence to scientifically authenticated truths (Fairhead & Leach, 2003) and investigating the authoritative representations of 'evidence-based knowledge' in planning for policy (Juntti et al., 2009; Eyben, 2013; Saltelli & Giampietro, 2017). Others have probed the ways that science makes nature understandable and amenable to control and management (Castree & Braun, 2001; Loftus, 2009; Taylor, 2014); exposing the links between science, states and markets with emphasis on the growing neo-liberalisation of these institutions (Goldman et al., 2011). Appreciating the wider implications of scientific authority means recognising the epistemological foundations of the technical and engineering sciences with their exclusions of power and interpretive dimensions of the use of resources and its place in social life (Zwarteveen, 2017). Critical perspectives on the dominant reliance upon technical fixes that address specific maladies (e.g. climate change) are mismatched with grounded experiences and needs, which themselves are shaped in large part by other political economic changes affecting communities, yet which lie outside the purview of socio-technical approaches (Nightingale et al., 2019). Such debates overall highlight that environmental science and policies are never free from their political contexts (Robbins, 2012). Feminist STS, for its part, fundamentally asserts that science is far from being impartial, highlighting its 'situatedness' (Haraway, 1988), and exploring ways to make this transparent: for what purposes, for whose interests, and how (gendered) knowers are themselves situated in contested spaces of power and resources?

Feminist epistemologies politicise knowledge production, but also necessarily focus on the structural and embodied forms of power that shape socio-envrionments, and that necessitate analysis and critique (Nightingale et al., 2019). This may be at odds with the requirements of environment and development, where the emphasis is on techno-fixes and managerialist appoaches. Here, the critical political edge of feminist knowledge becomes a liability, sometimes labelled as 'too complex or academic' (Ferguson, 2015). The kind of knowledge required from gender experts in technical settings includes compelling evidence to show that women's inclusion in environmental programmes will lead to favourable outcomes, among them, *better* climate adaptation and mitigation, water and sanitation management, disaster recovery and reconstruction and to some extent, better 'green' investments (Pearl-Martinez, 2014; Coleman & Mwangi, 2013; Harper et al., 2013; Harris & Abbott, 2018). This knowledge can leverage authority especially when it marshals evidence through the clinical objectivity of quantitative data. Technical organisations have their own inertia of doing things as defined by their knowledge practices and preferences: institutional rules, structures, practices, and the identities of the agents which animate them may continue to be primarily iterated to the political and social interests which institutions are designed to promote in the first place (Goetz, 1995: 2). This is also reflected in the revival and growing popularity of the use of economic cost–benefit analyses to justify the necessity for gender programmes in environmental organisations and in finance ministries. However, whether these organisations indeed accommodate gender programmes on these terms remains an under-researched area.

A popular example of this analysis is from a FAO publication circulating the notion that if women were more productive, the world would be less hungry: 'if women's access to productive resources were the same as men's, this would increase farm production yields by 20–30%' (FAO, 2011: 5). If left on its own, this analysis may inadvertently perpetuate gender stereotypes and the unequal reproductive food provision roles assigned to women, mobilise women as a reserve army of often undervalued labour without addressing their well-being interests and rights, and pass on to women the tasks of mitigating food insecurity that deflect from addressing its wider political economic causes and origins. As we write, climate change programmes and organisations are also trying to calculate the benefits of women's involvement in adaptation and mitigation measures. They highlight the severity of gender-specific impacts of climate change, anchored in the idea of climate change as an external, natural threat or hazard, sidelining its political economic nature that builds on the persistent economic dependence on fossil fuels. Accounting for gender impacts makes for a compelling case to include gender in climate change agendas, thus inviting planned interventions that seek to technically cushion and ameliorate these impacts without sufficiently addressing their fundamental causes (MacGregor, 2010; Tschakert, 2012).

Gender-informed evidence in the form of stylised facts, or short punchy messages with seductive statistics, is mobilised in strategic ways using the necessary vocabularies for crossing technical and social science divides (Cornwall et al., 2007). While these actions and strategies may indeed help form coalitions and play to a mixed audience of technical and business specialists making the gender agenda more palatable, recent work in FPE, feminist STS and decolonial thinking directs us to greater reflexivity and accountability with respect to the ontologies that are mobilised in environment and development work (Lamb, 2018; Clement et al., 2019). Specialists working on natural resource management historically align their work with the Enlightenment goal of intervening in the affairs of nature so that it is made useful and productive for human populations through science and technology. While almost indomitable, this ontology of productivism can be altered to more centrally position a socio-ecological justice framing for the well-being of human and non-human species. As calls for placing equity and a more politicised vision of sustainability at the analytical and policy center stage grow louder (e.g. Leach et al., 2018), specific types of knowledge need to be deconstructed by asking: knowledge for whom, knowledge for what? For example, Tuana (2013) examined an Inter-governmental Panel on Climate Change (IPCC) model of climate change risks and raised questions about the criteria being used to designate risk, which seemed to favour potential risks only for industrialised countries. Ultimately, by utilising such criteria, she asks whose interests are being favoured by the model and its data, and whether the well-being of those least responsible for but most affected by climate change is excluded from consideration.

Technical specialists are not required to see or understand the social realities within which their solutions and interventions are embedded (Zwarteveen, 2017). This reflects how the academic sciences have historically evolved in fragmented ways, obscuring the fact that knowledge is plural; however some knowledges are vested more authority than others. Researchers using multiple (e.g., positivist and non-positivist) epistemologies as a way to bridge knowledges are able to interrogate the silences between them and arrive at more amplified and less bounded understandings of socio-technical, environmental and gendered realities (Nightingale, 2003; Nightingale, 2016; Nightingale et al., 2019). From another vantage point, Harris (2015: 158), in her study of the multiple uses of water, acknowledges the importance of recognising 'everyday, embodied and emotional relations of people (and women) to resources and "natures"'. This is respect and recognition for 'local, historically-embedded and culturally-grounded forms of knowledge' (Leach et al., 2018: 8). By taking a step back and reflecting on their work, the contributors in this book are able to situate their mandate for transformative change within the landscape of authoritative knowledge in their own organisational context. Some of the contributors describe how, in navigating these different coalitions of knowledge, they are able to lever small openings for change. These brief excursions into the politics of knowledge can also lead to new ways of conducting gender analyses while realising epistemic justice (Aragon, 2019) and working with plural forms of knowledge beyond positivist science.

5. Collaborative reflexivity: our process and contributors

Our process for this book is built around generating knowledge collaboratively and relationally through reflexive conversation and embodied theorising. We invoke what Icaza and Vázquez (2016) describe as thinking from embodied experience and from the body. This involves us talking, thinking and writing through reflection on the experiences of our contributors. These are all people working in organisations that are variously positioned as sites of interaction for the central task of producing environmental knowledge and shaping policy norms. Reflexivity – conscious self-reflection on the implications of one's situatedness within different axes of power, privilege and affective relations – is an important element of feminist research and practice. A reflexive process underpins many of the contributions to ongoing debates around gender mainstreaming and gender expertise more broadly (Eyben, 2010). We have sought to work from a dialogic approach to reflexivity: using conversations to begin a process of reflexive analysis (Nagar, 2014), and from this we seek to capture ideas as they emerge from our co-learning encounters (Icaza, 2015; Browne et al., 2017). From these reflexive discussions, our writing strategy is to build engagements with wider theoretical and contextual debates. As Browne, et al. (2017) describe in their collaborative work, we have not always reached agreement on our interpretations, analyses and arguments: rather than smooth these over, we leave them in the text. In our choices around wider theoretical and contextual engagements, we are unable to decentre our editorial authority from the text entirely. Instead, we aim to be transparent about where authorial centering has taken place.

Our starting point was to develop our collaboration out of our professional 'rooted networks' (Rocheleau, 2015; Cantor et al., 2018), acknowledging that 'networks' necessarily involve exclusions and partial perspectives (Haraway, 1988). Each chapter is based on a series of conversations with contributors – writing and thinking as co-authors – who have a level of professional engagement with gender work and gender mainstreaming within technical environmental contexts. Each of us (editors and co-authors) is variously situated and privileged in terms of gender, race, career stage, coloniality and proximity to power within and outside our respective organisations. We have sought to avoid foreclosing around simple positional binaries in our editorial decisions, and instead our process (and the reflections of our contributors) is informed by 'intersectionality' where gender, race and other axes of power operate in dynamic interplay (Lykke, 2010; Cho et al., 2013). We share the view that gender and other aspects of domination are mobilised and hold specific meanings in specific contexts (Baines, 2010; Vita et al., 2016; Cho et al., 2013). In development work, questions around the implications of global north–south relationships and racialised privilege and authority may be accentuated (and perhaps simplified) more than is the case in other fields of work (Mollett & Faria, 2013). Intersectionality informs our process but also the reflections of our contributors. We add to this through a reflexive emphasis on how our positionalities intersect with the epistemic hierarchies around science and other ways of knowing within our respective work contexts that shape the tensions of being change agents (on the one hand) and also status quo bureaucrats or professionals, on the other (Eyben & Turquet, 2013).

We developed this as a form of collaborative autoethnography (Chang et al., 2016), which began with us inviting each of our contributors to write some brief thoughts on their positionality and role as a 'gender expert' within their specific institutional context. From this, we developed some questions to guide our initial conversations, which we both participated in. A follow-up conversation then took place to explore specific themes in more depth. The conversations were then converted into polyvocal essays, considered through relevant literature where this provides emphasis or enables deeper reflection and analysis. This 'embodied theorising' elicited deliberations on a range of experiences around issues of power, knowledge, positionality and identity, and the forces and factors that invest individuals and institutions with epistemic authority or deprive them of it (Aragon, 2019). For the most part we followed Browne et al. (2017) in maintaining the voices within each chapter in order not to flatten out our differences. In the spirit of mentoring from the page, we asked our contributors to share encounters that brought small 'wins' or learning points from within their practice. As each chapter shows, the conversation zoomed in and out between the everyday politics of bringing gender analysis into science and environmental discourse and institutional practices, and wider deliberations on epistemologies and power. What emerges on the page in large part reflects the subjectivities of our authors (as scholar-activists, as development practitioners) and ourselves in our editorial capacity.

The book divides into three interrelated themes where co-authors and editors engage in conversations about different but sometimes overlapping dimensions of each theme, grounded primarily on experience and self-reflexive thoughts on feminist practice. Under the first theme, *the politics of identity and boundary marking*, Maeve Nightingale of the International Union of the Nature Conservancy (IUCN) and Joy Clancy of ENERGIA and the University of Twente narrate their own boundary crossings from marine biology and mechanical engineering to gender research and policy development. In separate chapters, they narrate episodes where they engage respectively with homogeneous audiences of fellow engineers and gender specialists, reflecting on the strategies and actions that blur disciplinary boundaries, as well as those that constantly create them. Joy and Maeve tell us that outcomes from these episodes vary from more openness towards gender issues to creating unhelpful divides between gender experts and non-experts that can also occur in feminist circles. Gordon Prain reflects on a long career as a senior social scientist at the CGIAR International Potato Center (CIP). Through it, he highlights his changing, multiple yet interlocking identities where epistemic hierarchies at work kept him vacillating between positions of privilege and disadvantage. Andreea R. Torre disusses gender expertise with researchers at the Stockholm Environment Institute (SEI), Ha Nguyen, Lisa Segnestam, Emily Ghosh, Laura Forni, Natalia Biskupska and Marisa Escobar, citing immediate differences and deeper commonalities between gender experts, aspiring gender experts, and female technical scientists within a multi-disciplinary organisation, showing how the 'gender expert' is a contested claim and a complicated construct. Clara Mi Young Park of the UN Food and Agriculture Organization (FAO) unpacks the situated character of gender mainstreaming in politically difficult contexts thus deconstructing imagined divides between activists and femocrats in mainstream multilateral organisations like the UN. Kate Lappin of Public Services International (PSI) and former director of the Asia Pacific Forum on Women,

Law and Development (APWLD), on the other hand, remarks that gender work nowadays is being done in a safe way through its career-orientated trajectories of professionalisation, whereas she thinks that gender inequality transformation instead rests on building a new feminist fossil fuel-free future by strong social movements.

In the book's second theme, *the politics of knowledge in environmental and development realms*, contributors like Nicoline de Haan of the International Livestock Research Institute (ILRI) engage with livestock breeders, puncturing small openings for dialogue and collaboration along the practical lines of livestock health by setting up multi-disciplinary teams with complementary roles to play. Carol Colfer, formerly of the Center for International Forestry Research (CIFOR) discusses the importance of contextual learning to challenge universalisations of gender in forest contexts and the transferability of ethnographic methods in developing knowledge for forest governance. Marlène Elias of Bioversity International engages with technical scientists by navigating technical knowledges and using moments in the field to open spaces for dialogue around gender. Annette Wallgren and Victor Tsang of the UN Environment Programme (UNEP) pull the UN's mandated levers for gender and social equity integration with UNEP's internal programmes and optimise them as spaces for engagement, which they anchor firmly in the inter-linkages between people, society and environment. Margreet Zwarteveen of IHE Delft Institute for Water Education and Seema Kulkarni of the Society for Promoting Participative Ecosystem Management (SOPPECOM) tell of their engagements with water engineers in imagining water happiness maps instead of water productivity maps and enabling spaces for more deliberative democratic action among Indian civil society groups in tense water negotiations at subnational scales. Hilde Jakobsen, Napapan der Kinderen and Maria Holtsberg who work for disaster risk reduction (DRR) organisations describe entrenched hegemonic masculinities and the inertia of 'command and control' in the continuum between emergency management, humanitarian aid, and DRR.

The final theme is on the *power of gender champions* that brings donor reflections to the book. They discuss and showcase the positioning of gender agendas in their relations with partners and internally within their organisations. Maria van Berlokom, Eva Johanssen, Orawan Raweekoon and AnnaKarin Norling of the Swedish International Development Cooperation Agency (Sida) share their observations about how their partners respond to gender mainstreaming, highlighting the politics of knowledge within their partner organisations but also reflecting on their proximities to the spaces they as donors wish to transform. Senior gender advisors Kai Spratt and Charles 'Will' Lewis II openly tell of the thinning resources and commitment to gender equality within USAID and point to how this may be embedded in the wider politics and changing narratives around development aid more broadly.

As editors, we situate ourselves thus: Bernadette (Babette) as a woman-of-colour academic originally from the global South; and Rebecca (Becky) as a white woman from the United Kingdom, writing from her position as an academic in a teaching-led university. We have different professional biographies: whereas

Rebecca is a university teacher in Brighton focusing on thinking through feminist political ecology largely in a classroom context, Bernadette's work as a gender specialist with the think-tank Stockholm Environment Institute in Bangkok brings her more directly into development practice and policy research. Our journeys to these respective positions include women's movement activism (in the Philippines and in the UK), practising gender expertise directly in development projects (largely in Southeast Asia), and assuming the role of educator (in a university context and in practitioner workshops). The contributors, and indeed each of us as editors, are also variously situated in relation to the category 'gender expert', with some resisting the category entirely. Indeed, this professional categorisation was one of the points from which we launched our discussions.

References

Ahlborg, H., & Nightingale, A. J. (2017). Theorizing power in political ecology: The where of power in resource governance projects. *Journal of Political Ecology*. https://www.researchgate.net/publication/320249414_Theorizing_power_in_political_ecology_the_where_of_power_in_resource_governance_projects

Aragon, C. (2019). Global gender justice and epistemic oppression: A response to an epistemic dilemma. *Feminist Philosophy Quarterly*, *5*(2). https://doi.org/10.5206/fpq/2019.2.7294

Armiero, M., Barca, S., & Velicu, I. (2019, October). Undisciplining political ecology: A minifesto. *Undisciplined Environments*. https://undisciplinedenvironments.org/2019/10/01/undisciplining-political-ecology-a-minifesto/

Arora-Jonsson, S. (2014). Forty years of gender research and environmental policy: Where do we stand? *Women's Studies International Forum*. https://doi.org/10.1016/j.wsif.2014.02.009

Arora-Jonsson, S., & Sijapati Basnett B. (2017). Disciplining gender in environmental organizations: The texts and practices of gender mainstreaming. *Gender, Work & Organization*, *25*(3), 309–325. https://doi.org/10.1111/gwao.12195

Bacchi, C. (2009). The issue of intentionality in frame theory: The need for reflexive framing. In M. Verloo, P. Meier, & E. Lombardo (Eds.), *The Discursive Politics of Gender Equality: Stretching, Bending and Policy-Making* (pp. 30–41). Routledge.

Baden, S., & Goetz, A. M. (1997). Who needs [sex] when you can have [gender]? Conflicting discourses on gender at Beijing. *Feminist Review*, *56*(1), 3–25. https://doi.org/10.1057/fr.1997.13

Baines, D. (2010). Gender mainstreaming in a development project: Intersectionality in a post-colonial un-doing? *Gender, Work & Organization*, *17*(2), 119–149. https://doi.org/10.1111/j.1468-0432.2009.00454.x

Bauchspies, W. K., & Bellacasa, M. P. de la. (2009). Feminist science and technology studies: A patchwork of moving subjectivities. An interview with Geoffrey Bowker, Sandra Harding, Anne Marie Mol, Susan Leigh Star and Banu Subramaniam. *Subjectivity*, *28*(1), 334–344. https://doi.org/10.1057/sub.2009.21

Bee, B. A. (2019). Gendered spaces of payment for environmental services: A critical look. *Geographical Review*, *109*(1), 87–107. https://doi.org/10.1111/gere.12292

Bradshaw, S. (2013). *Women's Role in Economic Development: Overcoming the Constraints*. http://unsdsn.org/resources/publications/womens-role-in-economic-development-overcoming-the-constraints/

Browne, K., Banerjea, N., McGlynn, N., B, S., Bakshi, L., Banerjee, R., & Biswas, R. (2017). Towards transnational feminist queer methodologies. *Gender, Place & Culture*, *24*(10), 1376–1397. https://doi.org/10.1080/0966369X.2017.1372374

Bustelo, M., Ferguson, L., & Forest, M. (Eds.). (2016). *The Politics of Feminist Knowledge Transfer: Gender Training and Gender Expertise*. Palgrave Macmillan.

Butler, J. (1990). *Gender Trouble: Feminism and the Subversion of Identity*. Routledge.

Butler, J. (2004). *Undoing Gender*. Routledge.

Caglar, G., Prügl, E., & Zwingel, S. (2013). Introducing feminist strategies in international governance. In G. Caglar, E. Prügl, & S. Zwingel (Eds.), *Feminist Strategies in International Governance* (pp. 1–18). Routledge.

Calkin, S. (2017). Disrupting disempowerment: Feminism, co-optation, and the privatised governance of gender and development. *New Formations*, *91*(91), 69–86. https://doi.org/10.3898/NEWF:91.04.2017

Cantor, A., Stoddard, E. A., Rocheleau, D., Brewer, J. F., Roth, R., Birkenholtz, T., Foo, K., & Nirmal, P. (2018). Putting rooted networks into practice. *ACME: An International Journal for Critical Geographies*. *17*(4), 958–987.

Castree, N., & Braun, B. (Eds.). (2001). *Social Nature: Theory, Practice and Politics*. Blackwell Publishing.

Chang, H., Ngunjiri, F. W., & Hernandez, K.-A. C. (2016). *Collaborative Autoethnography*. Routledge.

Chant, S., & Sweetman, C. (2012). Fixing women or fixing the world? 'Smart economics', efficiency approaches, and gender equality in development. *Gender & Development*, *20*(3), 517–529. https://doi.org/10.1080/13552074.2012.731812

Cho, S., Crenshaw, K. W., & McCall, L. (2013). Toward a field of intersectionality studies: Theory, applications, and praxis. *Signs: Journal of Women in Culture and Society*, *38*(4), 785–810. https://doi.org/10.1086/669608

Clement, F., Harcourt, W., Joshi, D., & Sato, C. (2019). Feminist political ecologies of the commons and commoning (Editorial to the Special Feature). *International Journal of the Commons*, *13*(1), 1. https://doi.org/10.18352/ijc.972

Cochrane, L., & Rao, N. (2019). Is the push for gender sensitive research advancing the SDG agenda of leaving no one behind? *Forum for Development Studies*, *46*(1), 45–65. https://doi.org/10.1080/08039410.2018.1427623

Coleman, E. A., & Mwangi, E. (2013). Women's participation in forest management: A cross-country analysis. *Global Environmental Change*, *12*(1), 192–205.

Cornwall, A. (2007a). Buzzwords and fuzzwords: Deconstructing development discourse. *Development in Practice*, *17*(4–5), 471–484. https://doi.org/10.1080/09614520701469302

Cornwall, A. (2007b). Revisiting the 'gender agenda'. IDS Bulletin, *38*(2), 69–78. https://doi.org/10.1111/j.1759-5436.2007.tb00353.x

Cornwall, A., Harrison, E., & Whitehead, A. (2004). Introduction: Repositioning feminisms in gender and development. IDS Bulletin, *35*(4), 1–10. https://doi.org/10.1111/j.1759-5436.2004.tb00149.x

Cornwall, A., Harrison, E., & Whitehead, A. (2007). Gender myths and feminist fables: The struggle for interpretive power in gender and development. *Development and Change*, *38*(1), 1–20. https://doi.org/10.1111/j.1467-7660.2007.00400.x

Cornwall, A., & Rivas, A.-M. (2015). From 'gender equality and 'women's empowerment' to global justice: Reclaiming a transformative agenda for gender and development. *Third World Quarterly*, *36*(2), 396–415. https://doi.org/10.1080/01436597.2015.1013341

Cullen, P., & Murphy, M. P. (2018). Leading the debate for the business case for gender equality, perilous for whom?: Leading debate for the BCGE. *Gender, Work & Organization, 25*(2), 110–126. https://doi.org/10.1111/gwao.12199

Davids, T., van Driel, F., & Parren, F. (2014). Feminist change revisited: Gender mainstreaming as slow revolution. *Journal of International Development, 26*(3), 396–408. https://doi.org/10.1002/jid.2945

de Jong, S. (2016). Mainstream(ing) has never run clean, perhaps never can: Gender in the main/stream of development. In W. Harcourt (Ed.), *The Palgrave Handbook of Gender and Development: Critical Engagements in Feminist Theory and Practice* (pp. 92–105). Palgrave Macmillan.

de Jong, S., & Kimm, S. (2017). The co-optation of feminisms: A research agenda. *International Feminist Journal of Politics, 19*(2), 185–200. https://doi.org/10.1080/14616742.2017.1299582

Dersnah, M. A. (2019). United Nations gender experts and the push to focus on conflict-related sexual violence. *European Journal of Politics and Gender, 2*(1), 41–56. https://doi.org/10.1332/251510819X15471289106031

Desai, M. (2007). The messy relationship between feminisms and globalizations. *Gender & Society, 21*(6), 797–803. https://doi.org/10.1177/0891243207309907

Di Chiro, G. (2016). *Environmental Justice and the Anthropocene Meme* (T. Gabrielson, C. Hall, J. M. Meyer, & D. Schlosberg, Eds.; Vol. 1). Oxford University Press. https://doi.org/10.1093/oxfordhb/9780199685271.013.18

Elmhirst, R. (2015). *Feminist Political Ecology* (A. Coles, L. C. Gray, & J. H. Momsen, Eds.; pp. 58–66). Routledge.

Eyben, R. (2010). Subversively accommodating: Feminist bureaucrats and gender mainstreaming. *IDS Bulletin*, 41(2). https://doi.org/10.1111/j.1759-5436.2010.00123.x

Eyben, R. (2013). Uncovering the politics of 'evidence'and 'results': A framing paper for development practitioners. *Prepared for the Politics of Evidence Conference in Brighton*, UK.

Eyben, R., & Turquet, L. (Eds.). (2013). *Feminists in Development Organizations: Change from the Margins*. Practical Action Publishing.

Fairhead, J., & Leach, M. (2003). *Science, Society and Power: Environmental Knowledge and Policy in West Africa and the Caribbean*. Cambridge University Press.

FAO (Ed.). (2011). *Women in Agriculture: Closing the Gender Gap for Development*. FAO.

Ferguson, L. (2015). "This is our gender person." The messy business of working as a gender expert in international development. *International Feminist Journal of Politics, 17*(3), 380–397. https://doi.org/10.1080/14616742.2014.918787

Ferguson, L. (2019). *Gender Training: A Transformative Tool for Gender Equality*. Springer Berlin Heidelberg.

Forsyth, T. (2011). Politicising environmental explanations: What can political ecology learn from sociology and philosophy of science? In M. Goldman, P. Nadasdy, & M. Turner (Eds.), *Knowing Nature: Conversations at the Intersection of Political Ecology and Science Studies* (pp. 31–46). University of Chicago Press.

Fraser, N. (2009). Feminism, capitalism and the cunning of history. *New Left Review, 56*(2), 97–117.

Gluckman, P. (2016). The science–policy interface. *Science, 353*(6303), 969–969. https://doi.org/10.1126/science.aai8837

Goetz, A. M. (1995). Institutionalising women's interests and accountability to women in development. *IDS Bulletin, 26*(3), 1–10.

Goldman, M., Nadasdy, P., & Turner, M. (Eds.). (2011). *Knowing Nature: Conversations at the Intersection of Political Ecology and Science Studies*. University of Chicago Press.

Haraway, D. (1988). Situated knowledges: The science question in feminism and the privilege of partial perspective. *Feminist Studies*, *14*(3), 579–599.

Haraway, D. J. (2016). *Staying with the Trouble: Making Kin in the Chthulucene*. Duke University Press.

Harcourt, W. (Ed.). (2016). *The Palgrave Handbook of Gender and Development: Critical Engagements in Feminist Theory and Practice*. Palgrave Macmillan.

Harcourt, W., & Nelson, I. L. (2015). *Practicing Feminist Political Ecologies: Moving Beyond the 'Green Economy'*. Zed Books.

Harding, S. G. (1986). *The Science Question in Feminism*. Cornell University Press.

Harper, S., Zeller, D., Hauzer, M., Pauly, D., & Sumaila, U.R. (2013). Women and fisheries: Contribution to food security and local economies. *Marine Policy*, *39*(C), 56–63. https://doi.org/10.1016/j.marpol.2012.10.018

Harris, L. M. (2015). Hegemonic waters and rethinking natures otherwise. In W. Harcourt & I. L. Nelson (Eds.), *Practicing Feminist Political Ecologies: Moving Beyond the 'Green Economy'* (pp. 157–181). Zed Books.

Harris, S., & Abbott, K. (2018). *Climate and Women: The Business Case for Action*. BSR. https://www.bsr.org/en/our-insights/report-view/climate-women-the-business-case-for-action

Hoard, S. (2015). *Gender Expertise in Public Policy towards a Theory of Policy Success*. Palgrave Macmillan.

Holvikivi, A. (2019). Gender experts and critical friends: Research in relations of proximity. *European Journal of Politics and Gender*, *2*(1), 131–147. https://doi.org/10.1332/251510819X15471289106068

Icaza, R. (2015). Testimony of a pilgrimage. (Un)learning and re-learning with the South. In Z. Arashiro & M. Barahona (Eds.), *Women in Academia Crossing North-South Borders: Gender, Race, and Displacement* (pp. 1–26). Lexington Books.

Icaza, R., & Vazquez, R. (2016). The coloniality of gender as a radical critique of developmentalism. In W. Harcourt (Ed.), *The Palgrave Handbook of Gender and Development: Critical Engagements in Feminist Theory and Practice* (pp. 62–73). Palgrave Macmillan.

Juntti, M., Russel, D., & Turnpenny, J. (2009). Evidence, politics and power in public policy for the environment. *Environmental Science & Policy*, *12*(3), 207–215. https://doi.org/10.1016/j.envsci.2008.12.007

Kothari, U. (2005). Authority and expertise: The professionalisation of international development and the ordering of dissent. *Antipode*, *37*(3), 425–446. https://doi.org/10.1111/j.0066-4812.2005.00505.x

Kunz, R., & Prügl, E. (2019). Introduction: Gender experts and gender expertise. *European Journal of Politics and Gender*, *2*(1), 3–21. https://doi.org/10.1332/251510819X15471289106077

Kunz, R., Prügl, E., & Thompson, H. (2019). Gender expertise in global governance: Contesting the boundaries of a field. *European Journal of Politics and Gender*, *2*(1), 23–40. https://doi.org/10.1332/251510819X15471289106112

Lamb, V. (2018). Who knows the river? Gender, expertise, and the politics of local ecological knowledge production of the Salween River, Thai-Myanmar border. *Gender, Place & Culture*, 1–16. https://doi.org/10.1080/0966369X.2018.1481018

Lanz, K., Prügl, E., & Gerber, J.-D. (2019). The poverty of neoliberalized feminism: Gender equality in a 'best practice' large-scale land investment in Ghana. *The Journal of Peasant Studies*, 1–19. https://doi.org/10.1080/03066150.2019.1602525

Saltelli, A., & Giampietro, M. (2017). What is wrong with evidence based policy, and how can it be improved? *Futures*, *91*, 62–71. https://doi.org/10.1016/j.futures.2016.11.012

Sundberg, J. (2017). Feminist Political Ecology. In D. B. Richardson (Ed.), *International Encyclopedia of Geography: People, the Earth, Environment and Technology*. Wiley-Blackwell & Association of American Geographers.

Taylor, M. (2014). *The Political Ecology of Climate Change Adaptation: Livelihoods, Agrarian Change and the Conflicts of Development*. Routledge.

Tschakert, P. (2012). From impacts to embodied experiences: Tracing political ecology in climate change research. *Geografisk Tidsskrift-Danish Journal of Geography*, *112*(2), 144–158. https://doi.org/10.1080/00167223.2012.741889

Tuana, N. (2013). Gendering climate knowledge for justice: Catalyzing a new research agenda. In M. Alston & K. Whittenbury (Eds.), *Research, Action and Policy: Addressing the Gendered Impacts of Climate Change* (pp. 17–31). Springer Netherlands.

Valdivia, G. (2015). Eco-governmentality. In T. A. Perreault, G. Bridge, & J. McCarthy (Eds.), *Routledge Handbook of Political Ecology* (pp. 467–480).

van Eerdewijk, A., & Davids, T. (2013). Escaping the Mythical Beast: Gender Mainstreaming Reconceptualised. *Journal of International Development*, *26*(3), 303–316. https://doi.org/10.1002/jid.2947

Verloo, M., Meier, P., & Lombardo, E. (2009). *The Discursive Politics of Gender Equality: Stretching, Bending and Policy-Making*. Routledge.

Verschuur, C., & Destremau, B. (2012). Féminismes décoloniaux, genre et développement: Histoire et récits des mouvements de femmes et des féminismes aux Suds. *Revue Tiers Monde*, *209*(1), 7. https://doi.org/10.3917/rtm.209.0007

Vita, L. D., Sciannamblo, M., & Viteritti, A. (2016). Re-thinking intersectionality through Science and Technology Studies: Trajectories of women in technoscientific fields. *Rassegna Italiana Di Sociologia*, *3*, 503–524. https://doi.org/10.1423/84375

Zwarteveen, M. (2017). Hydrocracies, engineers and power: Questioning masculinities in water. *Engineering Studies*, *9*(2), 78–94. https://doi.org/10.1080/19378629.2017.1358730

establish common ground and 'fit in'. Once she tells a technical audience that she has a PhD in engineering, she can feel that people in the room begin to relax because they feel that 'here's a person who understands me'.

Joy: When I was offered the professorial chair in the university, the Dean wanted the title to be 'professor in gender and energy' and I said no, I want it as 'energy and gender' because if I put it the other way around, the majority of technical people will switch off. And it's a much better, stronger way of engaging with people and getting people to listen to what I want them to hear. Whereas if I put it the other way around, all the converted like the Babettes and Beckys will listen, but you don't get the sceptics to be convinced to listen to you.

Outside the University, some of the organisations with which Joy now works (for example, Africa Biogas, HiVOS (People Unlimited) and SNV (Netherlands Development Cooperation) are heavily dominated by male engineers. She believes that her technical background enables her to mix more easily in these circles, giving her a way in for introducing the role of women and/or gender in energy advocacy and research.

Joy: I'm really quite pragmatic in the sense of not profiling myself too much as a gender expert because in some ways it becomes a barrier since there's a lot of misunderstanding about what gender is about. Many think it's all about counting women. A lot of gender mainstreaming merely involves counting bodies and is not about changing values regarding inequality. I'm also not a social scientist so I cannot say that I'm a gender professional. I don't have a theoretical background in gender and feminist studies. My feminism comes from my political activism.

Eyben and Turquet (2013) note that feminists working in development organisations constantly have to make a difficult balancing act, and this is similar to what Joy alludes to in discussing her university and energy sector experience. But for Joy, it is clear that we (i.e. those working as 'gender experts') need to be conscious of how others perceive us as gender professionals or feminists. Avoiding this label may be one way of preventing them from 'switching off' at the first instance as this does not help realise our feminist goals of equality.

Joy's experience also points to the importance of recognising the dynamic, intersecting and open-ended character of how expertise is enacted and recognised in energy and gender work. Taking a lead from science and technology studies (STS) scholars and sociologists, 'expertise is a constantly changing legitimation of actions', and attention should be directed to the 'material and discursive conditions under which expertise itself is created and disseminated into a network' (Azocar and Ferree, 2016: 1084; Latour, 1987). Put another way, whilst it is helpful to understand the context of engineering as an expert field in order to reflect on openings for tackling gender inequality, it is also crucial to recognise the diversity and fluidity of that context and its framings.

As was noted by a contributor to Eyben and Turquet (2013: 61), '...you can become so entrenched in your identity as an activist against a dominant structure that you don't see the opportunities. You're just painting an entire field of people of a particular color...maybe that's why a lot people are turned off by feminism, because it's become reified in some way'.

Joy: It is quite important not to homogenise the world outside the 'gender profession' as my experience has shown me that it's very diverse out there [a point made by Faulkner, 2009a:172].

Joy locates this diversity in her own professional network in the history of intermediate (appropriate) technology (Milner, 2017), which was ascendant at the time when Joy was undertaking her PhD (Practical Action, n.d.). The network of engineers she was working with were part of the environmental movement and supported the idea that the entry point of appropriate technology was not only about the technology, but about people as well. Thus, the dualism between the technical and the social, which some sociologists of engineering have pointed out, was less pronounced. Furthermore, Joy's reflection is that the professional network of alternative energy engineers working in development contexts has meant she has worked with men who were supportive of her work around social and gender issues, and less aggressively masculinist, which made navigating the field of energy and gender much easier. The advantage, she says, lies in having mentors and a support network in an environment where your views (on gender, for example) don't easily fit.

Relating with the unconvinced: the ENERGIA network

The formation of ENERGIA, a network linking people from different disciplinary and professional backgrounds who had a shared interest in addressing the gender, development and energy nexus, is illustrative of an effort to bridge an epistemological gap between social and technical sciences, materialising a particular form of networked expertise (Azocar and Ferree, 2016). Joy illustrates the gap that existed prior to the founding of ENERGIA by noting that at the United Nations Beijing World Women's Conference in 1995, access to energy was not seen as a basic need so there was no UN agenda for it, and instead, it was subsumed under 'environment': a context which positioned women as managers and users of natural resources such as fuel wood. A paper was published in the influential journal *Energy Policy* by the Elizabeth Cecelski, a sociologist, based on her experiences at the International Labour Organization, that looked beyond the woodfuel crisis and improved stoves to focus instead on a broad range of issues around women's work, energy pricing and access, transport and modern energy forms (Cecelski, 1995). However, Joy notes that policy makers were not picking up the connection between gender and development and demand-side issues in the energy sector, and this lack of awareness is what prompted her to bring her engineering expertise into collaboration with Elizabeth Cecelski's social science expertise. Along with others, they went on to found ENERGIA. They diagnosed the problem as one of a limited evidence base that would enable them to broaden the debate, and they set out to address this, whilst at the same time, creating a 'space' to engage with policy makers. Whilst ENERGIA has gone on to be a leading organisation for research and advocacy in this area, we turn our attention to the 'small wins' that come about through Joy's attempts at bridging between different communities: engineers, policy makers, academics, private sector energy companies, and academic 'gender work'.

From the start, the approach has been pragmatic, with an emphasis on working in partnership for women's empowerment, gender and energy research, and gender and energy advocacy. Whilst there is an international secretariat based in the Netherlands, the network links 38 partners from across 18 countries, predominantly, but not exclusively in the South, working on projects, programmes and policies that explicitly address gender and energy issues. ENERGIA also creates training modules and tools for the energy sector, for example, on gender audits, to connect local initiatives across different contexts. Joy notes that there are challenges of bringing a gender perspective into play whilst connecting and working across different types of expertise.

What have been the strategies that have led to successes in research and policy advocacy within ENERGIA? For a start, Joy points to the importance of 'finding the right ear'.

Joy: I recognise that I am using my privilege at times and try to avoid speaking on behalf of women in the South, and also of creating new 'gender myths'.

For this reason, Joy (and Elizabeth Cecelski) have withdrawn to an advisory capacity although of late they have both become active in ENERGIA's research programme. The network is dominated by the global South, with Sheila Oparaocha as International Coordinator and Programme Manager of ENERGIA. From her own experience with ENERGIA, Joy makes the point that it is important to avoid making assumptions in either direction when working across disciplines and professions within gender training and advocacy networks.

Joy: For example, with the Africa Biogas Partnership Programme, which is very engineering-led. Initially there was resistance from the engineers to train women as brick workers, because this involved a change in women's roles and gender norms. I interviewed people from SNV [the NGO leading the programme] who admitted being taken aback that once trained, women were able to do this work [and that this was acceptable]. I find avoiding assumptions does work. And because I understand how the system works, the engineers are more accepting.

I don't disguise the fact that I'm interested in gender, and I try to encourage the idea that gender is relational and that we shouldn't push men out of the picture. A piece of work that I did for the European Commission a number of years ago was to show to the men who were working in the Commission that if they actually were much more gender sensitive or gender aware in their working environment, then it would also be for their betterment in the long run. I also said the same thing on a project with the Botswana Power Corporation telling company employees that gender equality is not only good for women, but that gender transformation is about transforming everybody's lives. And I think that then gets a much better buy-in. I'm not saying though that by saying these, there is absolutely no resistance.

Joy notes that some of the best 'wins' come when there are more men making the case for gender equality in energy systems. Within ENERGIA, she notes the work of Louis Seck, Energy 4 Impact Senegal Country and Project Manager, and describes how his leadership as an advocate for gender and energy has meant there is much better gender inclusivity in Senegal's renewable energy sector than there might otherwise have been.

Finding a common language: equality as efficiency and the power of numbers

Joy's pragmatism also echoes an approach that is common among gender professionals and bureaucrats, whereby the case for gender equality is made through an argument of the gains this brings for the economy: higher economic growth, better corporate performance and achievement of development goals. A longstanding literature explores how this is partly a response to the constricted spaces within which feminists in organisations must operate; this situation creates an imperative to articulate gender equality through a language that is familiar and widely accepted (Razavi, 1997; Eyben, 2010). The seductions of this approach are not without problems, including the instrumentalisation of gender to serve ends that are far from a transformation of women's experience (Caglar et al., 2013; Sandler and Rao, 2012; Kabeer, 2016; Razavi, 2017). What does this look like in the energy sector? Joy agrees that the 'efficiency argument' is what gains acceptance and significant traction.

Babette: There seems to be more and more attention being given now to gender and energy issues compared to, let's say, 10 years ago. You've been in the sector for a fairly long time, Joy, isn't there more attention to energy justice, women as untapped talent, and making the business case out of gender equality in the energy sector?

Joy: The Sustainable Development Goals (SDGs) now offer a leverage point, since they have established an 'energy' goal. The World Bank's women's economic empowerment agenda also offers a leverage point. The role of the private sector in delivering development offers an entry point. When it comes to the energy sector, there are emerging energy business models for women. However, there is little independent evaluation of this work. A focus on the economic inclusion of women may not necessarily lead to their empowerment.

For example, empowering women to sell cook stoves. The evidence does show that women do benefit from earning an income. I'm not against that at all, but it's the kind of emphasis on where in the energy supply chain women should be or be best at. And I don't think women are necessarily better at selling things. I just think that this is the stereotypical Avon lady model of female entrepreneurship! [i.e. women selling to other women who are 'like them'] so it is the inclusion of women as salespeople.

But it is also a reproduction of gender roles, that assumes only men could do technology, and therefore, women would be better sales agents. It's true though that in some cultures where it's difficult for a female family member to enter public economic space and that becoming a sales agent might represent empowerment, but still, it reinforces the present structures that separate women from technology. And I don't think selling cook stoves is going to radicalise gender roles.

But Joy notes that empowerment and transformation do not automatically follow gender-energy interventions.

Joy: In an evaluation project that I was involved in I realised that feminists face the uncomfortable thought that other women don't necessarily want transformation. For example, Magi Matinga, in her thesis on South Africa, finds still large piles of fuel wood outside households despite easier access to LPG stoves (Matinga, 2010). And this is because, in that particular place, a woman is judged by the size of her firewood pile. She's not a good wife and mother if the pile is low. Change in gender relations take a long time. It's not quick.

I think also that renewable energy has brought some interesting things to do with the energy transformation, not only the physical way of generating energy through local energy initiatives. Energy cooperatives, for example, can open up a space for women. The engagement of women is still very small in the Netherlands. And one of the reasons is time poverty. It doesn't necessarily always mean that it has to be big companies producing energy. But then on the other hand, I do wonder if this also partly reduces the role of the state in providing for equality of services for its citizens. For instance, if I happen to live in a locality where I've got lots of enthusiastic engineers who can put up solar panels, then I'll be alright. But if I happen to be in an area with people with less access to higher education who have less access to things, even in Europe, then I do wonder about energy security. It's a very political thing (the role of the state in the provision of utilities).

Joy's reflections contribute to a wider questioning of directionality in the relationship between gender equality and economic growth. Kabeer (2016) shows that gender equality contributes to economic growth, for example, as gender parity in education can spur a diverse and vibrant labour market. However, the inverse is not true. Bringing in women to sell cookstoves in Senegal or responsibilising women in The Netherlands energy cooperatives will not in itself lead to gender equality.

Joy points out that there are several conditions that need to be met if women are to be empowered in the energy sector, including education, entitlement to own property in their own right, and access to credit. Social and cultural factors may hinder women's economic empowerment – for example, whether or not women are permitted by the existing social norms of a society to be employed outside the home or, if they are, whether they control the use and investment of any income they earn. In other words, it may be a dangerous assumption that women's economic empowerment will lead to gender equality. Indeed, enabling energy access may create only the conditions that enable women to perform their traditional roles as care givers more efficiently, and possibly with a greater degree of comfort, while doing little towards transforming gender inequality (Clancy, 2016).

Our conversation with Joy turns to a second strategy for finding a common language, that is, bridging social and technical knowledges with numbers, as quantitative data is used to convince, demonstrate impact and claim authority.

Joy: Well, I think that decision-makers are primarily convinced by numbers. But numbers, as we know, don't tell the whole story. And there is this very irritating bit of data that's come out with the Global Tracking Framework [which charts progress on achieving SDGs (see IEA, 2019)].

SDG 7 [To ensure access to affordable, reliable, sustainable and modern energy for all] has three targets: access to energy and clean cooking, renewable energy and energy efficiency. In other words, electricity access and surprise, surprise, cook stoves! This means applying a gender lens because of women's association with cookstoves and the importance of clean [i.e. good indoor air quality] cooking. One of the gender myths, of course, is that female-headed households are poorer. Therefore, they have less access to electricity because of the high connection cost. However, for about ten countries in the world, the data shows something completely different: the percentage of female-headed households with access to electricity is higher than male-headed households. Why? No one can answer that because the statistical data doesn't answer that sort of question.

So we're left with a blindspot. One can only uncover the causes if one does qualitative work and preferably ethnographic investigation. When I proposed a qualitative data approach to the World Bank while I was doing a consultancy with them, I was met by a kind of incredulity and questions around what a qualitative analysis would show and how would it be helpful. And I suggested to them, 'Look, you only need to ask one question to people. Do you feel since you have had access to electricity, or your cookstove, or whatever, do you feel that your life has improved or that your family's life has improved?' And it either has or it hasn't, and then you can judge whether what you're doing is a success or not. And if it's a success keep doing it, and if it isn't, then you've got to go and find out why. 'Ask one question', I said. 'People are limited to being only the judges of what is better for them.

> Because what's better for me is not necessarily better for you or for a million and one other people. It's my life that matters. It's my life that I am the best judge of'. So it is the nature of the evidence that decision-makers are prepared to accept that I find worrying.
>
> What to do? Well, I mean, part of our gender and energy research programme is to produce some of the numbers to back up what we are saying. I think this is also useful for us in advocacy mode. You also have to find the right pitch and to say it often enough to one audience.

Metrics and quantitative impacts of a problem have taken on epistemological authority in mainstream development, particularly within traditionally technical domains such as energy. Used as evidence, statistics serve a performative function in that they can constitute the very reality in which policy makers and professionals make their interventions. For example, researchers have shown how the narrow focus on numbers can erase 'broader trajectories of socio-environmental change' from the policy domain as our knowledge of climate change impacts are directed towards specific framings (Tuana, 2013; Nightingale et al., 2019). However, there may be instances where certain kinds of quantitative instruments offer a common language and from this, discursive opportunities for reflexivity and change (Verloo and Van der Vleuten, 2009). It is in this space that energy and gender advocacy finds some of its traction for making visible and leveraging the linkages between gender equality and successful achievement of energy sustainability.

In conclusion

Many of the issues raised in our conversations with Joy echo with the experiences of gender practitioners in development settings more broadly. Our discussions around the power of numbers, for example, and the frustrations that accompany a knowledge hierarchy that defines quantitative data as 'evidence' and qualitative data as 'anecdote' is one that would find resonance for many gender researchers, particularly for those attuned to feminist debates around science, objectivity and situatedness (Haraway, 1988). Our starting point, however, was the extent to which the energy sector, which is embedded in engineering practice and the market logics of resource allocation, creates distinctive challenges for addressing gender equality and for the application of gender expertise. Joy's decades-long experience in this field, as a qualified engineer and as a founding member of the influential international research and advocacy network ENERGIA, points to some of the 'wins' – spaces from which change can happen, that she has experienced over the years. She adopts a strategy akin to what Carol Bacchi refers to as strategic reflexivity: enabling her to 'capitalize on small margins for change' (2009: 29).

Three 'small margins' stand out as 'take home' messages from our conversation. First, a common language is made easier where there is some co-recognition of knowledge and expertise between the gender researcher and other expert stakeholders. Joy's background as a mechanical engineer enables her to find common ground with technical colleagues. Her own story of how she became attuned to gender differences and inequalities in the energy sector has been instructive for how she brings these concerns to potentially sceptical colleagues (gently and without recourse to university-based feminist theorising). Secondly, whilst historically engineering has been shaped through its embeddedness within corporate and military institutions (Mills et al., 2014), Joy's engagement with this technical field is instead situated within an alternative development paradigm of intermediate (appropriate) technology. Joy makes the point that this branch of engineering has been open to engaging with people, making the step towards embracing gender only a small one. The challenges of finding a common language are perhaps less with an epistemological gap with engineers of appropriate technology, and more with a context where the primary engagement is with policy makers working at scale and in the context of neoliberal development, with all its technologies of log frames and quantitative impact indicators. Thus, thirdly, Joy and her colleagues have attempted to address this challenge by creating an expert network encompassing a diversity of people, things and ideas in the form of ENERGIA – a network that produces knowledge on gender and energy (through relatively conventional mainstreaming tools such as gender audits) and makes it effective in the world through advocacy work.

However, as some of our other contributors have outlined, whilst the situated everyday practices that make up Joy's work with ENERGIA have opened up room for manoeuvre, by far the most important factor in enabling gender to be a key part of energy and development work is the convergence of the SDGs (and their global tracking) and the impetus for energy transformation to address climate change. The performativity of each of these makes it all the more important that the right kinds of gender work materialise, and that these do not go on to constitute another trap or burden for the disadvantaged.

References

Azocar, M. J., & Ferree, M. M. (2016). Engendering the sociology of expertise. *Sociology Compass*, *10*(12), 1079–1089. https://doi.org/10.1111/soc4.12438

Bacchi, C. (2009). The issue of intentionality in frame theory: The need for reflexive framing. In M. Verloo, P. Meier, & E. Lombardo (Eds.), *The Discursive Politics of Gender Equality: Stretching, Bending and Policy-Making* (pp. 30–41). London: Routledge.

Caglar, G., Prügl, E., & Zwingel, S. (2013). Introducing feminist strategies in international governance. In G. Caglar, E. Prügl, & S. Zwingel (Eds.), *Feminist Strategies in International Governance* (pp. 1–18). Oxon, UK: Routledge.

Cecelski, E. W. (1995). From Rio to Beijing. Engendering the energy debate. *Energy Policy*, *23*(6), 561.

Cecelski, E. (2004). *Re-thinking Gender and Energy: Old and New Directions*. ENERGIA/EASE.

Cecelski, E., & CRGGE, (the Collaborative Research Group on Gender and Energy). (2006). *From the Millennium Development Goals Towards a Gender-Sensitive Energy*

Policy Research and Practice: Empirical Evidence and Case Studies—Synthesis Report. The Hague: ENERGIA.

Cecelski, E., & Dutta, S. (2011). *Mainstreaming Gender in Energy Projects: A Practical Handbook. Resource Pack.* Retrieved from ENERGIA website: http://genderandenvironment.org/resource/mainstreaming-gender-in-energy-projects-a-practical-handbook-resource-pack/

Clancy, J. (2016). Transforming energy systems and gender roles and relations are interconnected. Inaugural lecture Joy Clancy. Available online from ENERGIA.ORG. https://www.energia.org/?s=Joy+Clancy

Clancy, J. S., Skutsch, M., & Batchelor, S. (2002). *The Gender-Energy-Poverty Nexus: Finding the Energy to Address Gender Concerns in Development* (No. DFID Project CNTR998521). Retrieved from DFID website: https://research.utwente.nl/en/publications/the-gender-energy-poverty-nexus-finding-the-energy-to-address-gen

Connell, R. (2016). Masculinities in global perspective: Hegemony, contestation, and changing structures of power. *Theory and Society*, *45*(4), 303–318.

Danielsen, K. (2012). *Gender Equality, Women's Rights and Access to Energy Services . An Inspiration Paper in the Run-up to Rio+20.* Amsterdam: Royal Tropical Institute (KIT).

de Groot, J., Mohlakoana, N., Knox A, & Bressers, H. (2017). Fuelling women's empowerment? An exploration of the linkages between gender, entrepreneurship and access to energy in the informal food sector. *Energy Research & Social Science*, *28*, 86–97

ENERGIA. (2015). Mainstreaming gender in energy projects. ENERGIA International Network on Gender and Sustainable Energy. http://energia.org/wpcontent/uploads/2015/02/01.-Mainstreaming_gender_in_energy_projects_A_practical_Hand_book.pdf

Ettinger, L., Conroy, N., & Barr II, W. (2019). What late-career and retired women engineers tell us: gender challenges in historical context. *Engineering Studies*, 11(3), 217–242

Eyben, R. (2010). Subversively accommodating: feminist bureaucrats and gender mainstreaming. *IDS Bulletin*, 41(2). https://doi.org/10.1111/j.1759-5436.2010.00123.x

Eyben, R., & Turquet, L. (Eds.). (2013). *Feminists in Development Organizations: Change from the Margins*. Rugby: Practical Action Publishing.

Faulkner, W. (2009a). Doing gender in engineering workplace cultures. I. Observations from the field. *Engineering Studies*, *1*(1), 3–18. https://doi.org/10.1080/19378620902721322

Faulkner, W. (2009b). Doing gender in engineering workplace cultures. II. Gender in/authenticity and the in/visibility paradox. *Engineering Studies*, *1*(3), 169–189. https://doi.org/10.1080/19378620903225059

Gay-Antaki, M. (2016). 'Now we have equality': A feminist political ecology analysis of carbon markets in Oaxaca, Mexico. *Journal of Latin American Geography*, *15*(3), 49–66.

Haraway, D. (1988). Situated knowledges: The science question in feminism and the privilege of partial perspective. *Feminist Studies*, *14*(3), 575–599.

IEA (International Energy Agency). (2019). *Tracking SDG7: The Energy Progress Report*. https://trackingsdg7.esmap.org/

Jasanoff, S. (2004). *States of Knowledge: The Co-Production of Science and Social Order*. Routledge.

Kabeer, N. (2016). Gender equality, economic growth, and women's agency: the 'endless variety' and 'monotonous similarity' of patriarchal constraints. *Feminist Economics*, *22*(1), 295–321.

Kelkar, G., & Nathan, D. (2005). *Gender Relations and the Energy Transition in Rural Asia*. Retrieved from Collaborative Research Group on Gender and Energy (CRGGE) with support from the ENERGIA International Network on Gender and Sustainable Energy website: https://www.gov.uk/dfid-research-outputs/gender-relations-and-the-energy-transition-in-rural-asia

Kooijman, A., Cloke, J., & Clancy, J. (2018). Needs, wants and values: Integrating gender with energy access. Briefing paper 3. Low Carbon Energy for Development Network and ENERGIA. December 2018. https://www.energia.org/cm2/wp-content/uploads/2019/03/Final-Gender_Low-Carbon-Energy-Network-Briefing-Paper-3-AW-web.pdf

Kronsell, A. (2011). Gendered practices in institutions of hegemonic masculinity. *International Feminist Journal of Politics*, *7*(2): 280–298.

Lambrou, Y., & Piana, G. (2006). *Energy and Gender Issues in Rural Sustainable Development*. Rome: FAO.

Latour, B. (1987). *Science in Action: How to Follow Scientists and Engineers through Society*. Boston, MA: Harvard University Press.

Leach, M., Scoones, I., & Stirling, A. (2010). *Dynamic Sustainabilities: Technology, Environment, Social Justice*. London: Earthscan.

Liebrand, J. (2014). *Masculinities among Irrigation Engineers and Water Professionals in Nepal* (Wageningen University). http://edepot.wur.nl/321002

Matinga, M. N. (2010). '*We Grow Up With It*': *An Ethnographic Study of the Experiences, Perceptions and Responses to the Health Impacts of Energy Acquisition and Use in Rural South Africa*. PhD Thesis, University of Twente, Enschede, The Netherlands.

Miller, C. A., Richter, J., & O'Leary, J. (2015). Socio-energy systems design: A policy framework for energy transitions. *Energy Research & Social Science*, *6*, 29–40.

Mills, J.E., Franzway, S., Gill, J. and Sharp, R. (2014). *Challenging Knowledge, Sex and Power: Gender, Work and Engineering*. London: Routledge.

Milner, T. (2017) *From Intermediate Technology to Technology Justice: The Knowledge Sharing Journey of Practical Action* https://www.theimpactinitiative.net/impact-lab/collection/intermediate-technology-technology-justice

Nelson, S., & Kuriakose, A. T. (2017). *Gender and Renewable Energy: Entry Points for Women's Livelihoods and Employment*. Climate Investment Funds.

Nightingale, A. J., Eriksen, S., Taylor, M., Forsyth, T., Pelling, M., Newsham, A., … Whitfield, S. (2019). Beyond technical fixes: climate solutions and the great derangement. *Climate and Development*, 1–10. https://doi.org/10.1080/17565529.2019.1624495

Pearl-Martinez, R., & Stephens, J. C. (2016). Toward a gender diverse workforce in the renewable energy transition. *Sustainability: Science, Practice, & Policy*, *12*(1). http://search.proquest.com/openview/33f4fc69af7b9886ecea63b4565af3d1/1?pq-origsite=gscholar&cbl=136137

Practical Action (n.d.). Who we are and what we do. https://practicalaction.org/who-we-are/

Rathgeber, E. M. (1990). WID, WAD, GAD: trends in research and practice. *The Journal of Developing Areas*, *24*(4), 489–502.

Razavi, S. (1997). Fitting gender into development institutions. *World Development*, *25*(7), 1111–1125. https://doi.org/10.1016/S0305-750X(97)00023-5

Razavi, S. (2017). Revisiting equity and efficiency arguments for gender equality: A principled but pragmatic approach. *Canadian Journal of Development Studies/Revue Canadienne d'études du Développement*, *38*(4), 558–563. https://doi.org/10.1080/02255189.2017.1376624

Rojas, A. V., Schmitt, F. M., & Aguilar, L. (2014). *Guidelines on Renewable Energy Technologies for Women in Rural and Informal Urban Areas*. Retrieved from IUCN and ENERGIA website: http://genderandenvironment.org/resource/guidelines-on-renewable-energy-technologies-for-women-in-rural-and-informal-urban-areas/

Saks, M. (2012). Defining a profession: The role of knowledge and expertise. *Professions and Professionalism*, *2*(1). https://doi.org/10.7577/pp.v2i1.151

Sandler, J., & Rao, A. (2012). *Strategies of Feminist Bureaucrats: United Nations Experiences* (Working Paper No. 397). Sussex, UK: Institute of Development Studies.

Tuana, N. (2013). Gendering climate knowledge for justice: catalyzing a new research agenda. In M. Alston & K. Whittenbury (Eds.), *Research, Action and Policy: Addressing the Gendered Impacts of Climate Change* (pp. 17–31). Dordrecht: Springer Netherlands.

Verloo, M., & Van der Vleuten, A. (2009). The discursive logic of ranking and benchmarking Understanding gender equality measures in the European Union. In P. Meier, E. Lombardo, & M. Verloo (Eds.), *The Discursive Politics of Gender Equality: Stretching, Bending and Policy-Making* (pp. 169–185). London: Routledge.

2 Is epistemic authority masculine?

Reflections on gender, status and knowledge in international agricultural research and development

By Bernadette P. Resurrección and Rebecca Elmhirst
In conversation with Gordon Prain

A defining feature of feminist political ecology (FPE) is that it interrogates power around knowledge production and the positionality of the knower in a world divided in intersecting ways along lines of gender, class, social status, and age. It was in this same spirit that feminist epistemologist Lorraine Code, then a young scholar in 1981, was compelled to pose the question, 'Is the sex of the knower epistemologically significant?' Indeed, FPE owes a huge debt to feminist epistemology in positing that the bearer of knowledge – the knower – is embodied, situated and embedded as opposed to an unmediated, faceless, context-free and bodiless knower standing 'from an aloof and elevated position of surveillance' (which Haraway (1988) refers to as the 'god trick' or the view from nowhere). In short, one's social location (gender being one such dimension) both shapes and limits one's knowing (Lykke, 2010; Haraway, 1988).

But what about knowledge itself – are some types of knowledge more hegemonic, or considered superior? This question resonates with the growing numbers of women gender experts in science, environment and technology organisations, where they face often implicit social hierarchies that intersect what they know with who they are. Whose knowledge counts? What role, if any, does professional identity, and the composition of one's professional field, play with regard to one's gender, ethnicity, class, age and the reception of one's work? Are there straightforward associations between the epistemic authority of techno-scientific knowledge with masculinity as contemporary scholars sometimes claim? (MacGregor, 2010; Sismondo, 2004). The insights from a long career history can give us clues to the relationship between knowledge, identity and power.

Gordon Prain, male senior researcher in the CGIAR (Consultative Group on International Agricultural Research) global agricultural innovation network, is a social scientist and gender expert who looks back at 30 years of engagement in various positions, mostly within the International Potato Center (CIP). CIP is an organisation dedicated to improving the livelihoods of low income farming families through research and development of roots and tuber crops. Through social science research within CIP and through his later membership of the Participatory

Research and Gender Analysis (PRGA) Program, a cross-centre program of the CGIAR, he devoted time with agricultural research professionals to explore with them who 'the farmer' was. He eventually discovered that they generally understood 'the farmer' as the male farmer. His recognition of this biased view was the beginning of a long career as a gender expert as well as social scientist in the CGIAR system.

Through first-hand experience, Gordon narrates to us how a global research-for-development organisation began to accommodate gender dimensions in its strategic programming and research activities. The landmark emergence of participatory development in the 1980s threw open the gates for social issues and questions to inhabit future agricultural research. Under the aegis of participatory development, the entry of gender research catalysed debates on whether it was enough to include women in agricultural development processes; or whether the kind of development that was enveloping the rural world was widening gender and wealth gaps, and therefore it was instead the locus of needed change and intervention.

Participatory development unpacks 'the farmer'

Gordon: I am recently officially retired, but I still work as a consultant. I've been working for the CGIAR for more than 30 years. I'm an anthropologist and I've been leading social and nutrition sciences at CIP in recent years. For 10 years I was also global coordinator of a CGIAR system-wide initiative called Urban Harvest, which undertook research and development initiatives on urban and peri-urban agriculture, and urban environmental issues. My gender professionalism has come through applied social science work, I would say, over a long period of time. I think a lot of people of my generation working on gender were strongly influenced by the emergence of participatory research.

Becky: I think that it's interesting how participatory research has been an entry point for a number of people into this world of gender. It's interesting what's coming through also from this is how this approach opened a lot of doors for social questions and that was almost the more transformative moment in agricultural research than what we're seeing now with a recent push for gender analysis.

Gordon: I coordinated an R&D network called UPWARD for seven years in Asia. The acronym stands for Users' Perspectives with Agricultural Research and Development and it focused attention on the end-users of agricultural innovations, in other words farming communities and farming households. And, of course, when you start looking at those users' perspectives or points of view, you have to look at both men's and women's perspectives. And so that experience was quite influential in professionalising, if you like, my own sort of approach to gender. That work in Asia with users' perspectives increased the attention towards gender. We were listening to men and women's voices, differentiating men and women's needs, as well as supporting technology adaptation that was appropriate for both men and women, and the overall attention we were paying to gender percolated through to the senior management within CIP and to other scientists, but not yet then in a very well-articulated way. In other words, at that stage, we were articulating the importance of listening to users, and within that, we were pointing out that users were not generic but different since in international and national agricultural and environmental research scientists mostly simply referred to 'the farmer' in aggregate terms. Tacitly, it was a male farmer. We were pointing out that you have to disaggregate, you have to understand women and men's different needs. So, I think it had an influence, but it wasn't a sufficiently strong influence in moving the Center (CIP) forward in terms of its own institutionalisation of gender issues, but it was a fair start. But since 2000, in cross-Center initiatives and programs like Participatory Research and Gender Analysis (PRGA), Urban Harvest, the new CGIAR Research Programs (CRPs), especially on Roots, Tubers and Bananas (RTB), and most recently through GENNOVATE, a cross-CRP project seeking to enable gender equality in agricultural and environmental innovation, I had a much more active and direct role in articulating gender as a policy and organisational principle since I was advising or leading teams of researchers and gender professionals across a number of CGIAR Centres to conduct gender research and develop gender strategies within research programs.

Certainly, the institutional space for these recent opportunities in gender research and policy change has been strongly influenced by changes in donor priorities. For at least the last 15 years there has been growing pressure on leaders and senior management across the CGIAR to take up gender more seriously.

From the early 1980s, participatory rural development was influential in large part due to a growing discontent with how development initiatives in the global South were being implemented, especially the 'top down' 'transfer of technology (TOT) model underpinning actions that despite the Green Revolution left millions of the most marginalised communities still in poverty (Biggs, 1990; Horton et al., 1989). These problems of mainstream economic development led to the nurturing of more inclusive and less 'top-down' approaches. Robert Chambers was an early pioneer on this. In his landmark book, *Rural Development: Putting the Last First* (1983), he spelt out the disconnect between development professionals and the poor, the need for participatory development. The recognition of the power dimensions of 'rural poverty unperceived', as Chamber's opening chapter is called, was fertile ground for recognising gender inequality in development: 'In crop research, priority, prestige and promotion have gone with work on crops for export, grown usually by plantations, large farmers, the better off small farmers, and the men of the household rather than the women' (Chambers, 1983: 77). Henceforth, debates followed on whether it was a matter for women to be included in development (WID or Women-in-Development framework), or whether development itself was a problematic site of unequal gender and power relations (GAD, or Gender and Development framework) in the first place. GAD demanded a revamp and overturning of the dominant development model that caused disadvantage to women on many fronts (Rathgeber, 1990). This distinction between WID and GAD was important since each led to a distinct analytical and implementation practice. Today, we are witnessing an incipient resurgence of this debate as socialist feminist scholars are calling for a re-examination of the dominant model of development that is premised on neoliberal economic growth. Those taking this position are critical of the neoliberal tendencies of influential strains of feminism that stress the economic inclusion of women rather than change of the model of 'development' itself (Funk, 2012; Prügl, 2015; Fraser, 2009). In research organisations where gender is a consideration, there were moments that this debate had real impacts on how gender integration in research was to proceed, and in some instances, the debate stalled the process. Gordon recalls one moment of stasis.

Gordon: There was an internally commissioned review of the Participatory Research and Gender Analysis (PRGA) program of the CGIAR in 2000 of which I was the Chair. We definitely found that the participatory components of PRGA were way ahead and more successfully implemented than the gender analysis component. Whereas the participatory research component helped mainstream participatory plant breeding and achieved a wide dissemination of participatory tools in international and national research organisations, the gender analysis component was rather bogged down in theoretical disputes, especially around women in development versus gender (and development) analysis. At the time we did the review there was a limited application of gender analysis or the use of gender tools in research activities. As a result, we were critical of that weakness in the program, recognising that participatory development as a research strategy did not have a strong gender analysis component. At that stage PRGA wasn't able to articulate – let's say in a clear, conceptual way – the gender analysis aspects. But in the practice of participatory research such as in participatory plant breeding, people were paying attention to what women and men were asking for in terms of varieties, in terms of traits, and so on. So later in 2005–2006, more attention was paid to gender analysis.

The landing of gender issues and analyses was not smooth in the world's largest international agricultural research network. It collided with the organisation's traditional technical focus and came at a time when ontological debates within gender and development were being sorted out. Nevertheless, participatory development ushered in the social dimensions of agricultural development. On the ground, Gordon witnessed how technical and social professionals began to seek ways to relate to each other. In the next section, we pay more attention to how professional identities and knowledge are inter-connected through relations of power.

Situated identities

The evolution of gender knowledge in the CGIAR shaped how Gordon charted his career as a gender professional in rural and agricultural development. The growing knowledge of and dedication to research on gender stretched and bended alongside Gordon's own rise to seniority in a context where overall, gender studies continued to remain less important than other types of knowledge.

Babette: Many agricultural development organisations are male-dominated but the gender experts I know in those organisations are mostly women. It's not all the time that a man gets to be a gender professional, and I'm just curious how you eventually came to that point, Gordon, in that your colleagues, or the organisation you worked for, actually came to recognise you as a gender professional. What sort of led you to that point? And how has it been received?

Gordon: I think people have come to see me as a kind of spokesperson for the intended beneficiaries of the work going on in the Centre, and in the broader CG [short name for the CGIAR system[, in a way. There was an expectation for me to constantly remind people developing technologies that their work should be socially relevant. For example, have they asked the right questions about the people they work with? And then, in that, have they listened to the different demands coming from men and women? As a social scientist who's spent a lot of years in the CGIAR, I heard a lot of moaning and groaning [laughter] about being the underdog from other social scientists. Constantly put down, constantly not listened to, and very much of a second-class citizen. And that's probably the most common story among social scientists.

But as a man, I've not been made to feel that. Yes, as a young anthropologist I was ignored by the senior bio-physical scientists, but I've not suffered. I've not been marginalised to the extent that women gender researchers have been marginalised. I think that's fair to say, that it's been very difficult often for women social scientists to be heard in a bio-physical science environment. I've always argued very strongly, and I still do with my colleagues who are women gender professionals, that the first kind of step along the road to being heard is to understand as much as you can of the technical side of things. I really emphasise that. There should be at least understanding on what breeders are doing, and why they're doing what they're doing. What are the concerns of people involved in integrated pest management? What is involved with the whole issue of seed? Why is seed so important? These things, for a gender professional, are extremely important to understand to enable them to engage technical people in conversations about those issues. I was so heavily involved in issues of women and men's on-farm conservation of crop genetic diversity at one stage of my career that I was taken for an enlightened geneticist! With this understanding of the technical dimension, we can then start to point out how technologies can do harm or good and anticipate possible gender traps. If gender professionals come from a position of complete innocence around those technical issues, it then makes the job extremely difficult.

So, what I'm trying to say is that, yes, I've had an easier time as a man but at the same time, I was engaged in a field that was not as important as other fields. But you have to understand where people in these fields are coming from and then respond in as strong a way as possible to their ideas and present the relevance of gender to their fields. I think that this kind of conversation is now taking place. And I now feel very comfortable putting gender right at the top of the agenda of agricultural research. It's not always been like that.

Becky: You mentioned about having an easier time as a man. Do you mean that you are being taken seriously in a kind of interpersonal context? So, when you walk into a room, do you think people are more likely to listen to you than, say, a young woman with a background on gender? Is that what you mean?

Gordon: I think it is that. But what I'm not sure about, and this is where one has to be very careful is that in recent years, I've had a senior position. So, it is difficult to disentangle gender from rank and position. Or from status, if you like. Doing gender work within the new CGIAR Research Programs, I feel as though people have really listened to me because I was one of the very few senior researchers who actually was heavily involved in gender.

Babette: It's helpful for gender, but I was wondering if you walked into a room and there was a woman, let's say, in a meeting, and she was not a gender specialist or professional, but rather, she was a technical person, shall we say, a geneticist or a crop scientist. Who would be more listened to?

Gordon: Well, again [laughter], if it was in 1988 or 1987, for sure that she would have had the much more important voice and people would have listened to her much more. I worked with a woman geneticist, a breeder, in the 1980s in Latin America and we were working together on developing new varieties and testing them with farming communities, men and women. I think it was a very powerful combination in a way. At that time, I was very junior, but nevertheless, it was the voice of a male anthropologist who was pushing to listen to men and women about which varieties were important. The woman breeder listened to me about how to adjust her work and how to consult with different household members about the development of these new varieties.

In 2010, as a more senior social scientist and male gender researcher, I would have been listened to as much as a woman breeder. Whether I would have been **heard** is another matter and remains a question. Even in 2010, I doubt that a young woman gender specialist would have been listened to as much as a woman breeder. We are dealing with multiple identities linked to gender, age and disciplinary status: intersectional power differences which we need to disentangle.

From his recollections as a young anthropologist negotiating a social research agenda with plant breeders, Gordon's stature changed as he gradually attained seniority. His positionality as a western white man, senior in the profession, and a social scientist in a technical research organisation indicates to us that power, identity and knowledge are co-created. At some critical junctures in his organisation's history, his privileged positionality has been able to temper internal resistance to the entry of non-technical epistemological fields such as gender studies. However, this has not fully led to equal epistemic status with the technical sciences, which then dents Gordon's privileged position as a senior scientist. Whereas most studies explore multiple oppressions, Gordon's story emphasises the peculiar situated character of intersectionality: that of experiencing simultaneous marginalisation and privilege through fluid identities and interactions shaped by the dynamics of epistemic power and authority in this specific work environment. Intersectionality scholars further warn us not to see multiple and interlocking identities of gender, class, age, race as static and caught in a gridlock. By empirically recognising them, and 'how they work' and change over time and context, they allow us to unpack institutional power and epistemic dynamics that have enabled them to take shape and emerge at critical moments (Cho et al., 2013; Garry, 2011; Vita et al., 2016).

Gordon has had the privileged experience of looking back at a long career history interlaced with shifting positionalities that enabled positive interactions with technical specialists. Younger gender professionals may have slightly different intersectional experiences and trajectories as they engage in gender mainstreaming research for agricultural development.

Interactions with technical specialists and young gender professionals

Becky: Were there positive collaborative experiences with technical scientists?

Gordon: Cases of collaboration were very specific. I want to emphasise that within the CGIAR, CIP has been one of the most successful International Research Centers in pioneering collaboration between technical and social scientists, especially in relation to participatory 'bottom up' field research. Collaborative experiences were often very intensive and successful at the level of the individual in the sense that the individual technical person took on and understood the relevance and importance of engaging directly with farmers and of understanding gender. But there were few outcomes beyond that individual. I think that was the biggest challenge of that time, moving from individual to institutional change.

I worked with geneticists and plant breeders in Peru and Indonesia who were interested in conserving crop genetic diversity and using that diversity to develop new varieties of potato and sweet potato. In the Peruvian experience the plant breeder, who already had exposure to social sciences through her siblings, was very open to collaboration in order to develop a new variety with a certain type of resistance to a soil pest seriously affecting high altitude potato cultivation, on which communities depended for food. And so, I encouraged her and introduced her to a methodology of how to tap into the voices of women and men farmers about the kind of preferences and concerns they had around potato varieties, the things that were causing them problems, and so on. We had a systematic process of evaluation involving men and women. And from that breeding work, I think more people – both women and men benefited. And in the end, she actually released a variety with a woman's name [laughter]. I think she really took on the importance of understanding differences and preferences between men and women farmers, and which she incorporated into her work. But it didn't go any further than that within CIP's plant breeding program.

In West Papua, in the Indonesian half of New Guinea, I worked with a geneticist to collect and document sweet potato varieties. This is a secondary center of diversity of the crop, the main food staple and central to local culture. The documentation process involved talking to men and women about the large number of varieties they cultivated and used. And it clearly came out that there were very different preoccupations in terms of the types of varieties and the qualities that were important to women and men. This geneticist was not at all sensitive to gender issues before we started working. We had a very intensive period of plant collection in New Guinea. And that very down-to-earth experience had an impact on him. But again, it didn't seem to go any further than that.

At a more programmatic level, I was responsible for a program related to urban agriculture. We had a strong focus on gender in that program, including the co-production with a Dutch foundation of a volume of methods and case studies called 'Women Feeding Cities' (Hovorka et al., 2009). We wanted to move from talking to action. We felt one way to do that was to actually get into a research partnership with a technical seed specialist and to work together on data collection. It was clear that there was a need to undertake joint action between a gender specialist and a technical scientist. We also had to bring our own financial resources. It's easier for researchers to say that we should be paying attention to men and women than to actually allocate resources to do that.

So, the work was completed with interesting results that were published as a journal article (Njenga et al., 2011). But when it came to present the results of this work in public – to a CIP science meeting – only the technical aspects of what had been achieved were presented, not the gender dimensions.

Gordon: Coming back to my earlier discussion of the CGIAR System-wide Initiative on Participatory Research and Gender Analysis (PRGA), I want to comment on how the program tried, early on, to strengthen gender analysis within the different International Agricultural Research Centers which participated. One strategy, which was actually inherited from the earlier gender program within the CGIAR which was responsible for both gender in the workplace as well as gender in research, involved naming 'Center Liaisons', predominantly young and often female researchers tasked to push the gender agenda within the research programs of those Centers. But to be honest, those positions were fairly invisible. They had no institutional recognition, no funding and as the 2000 internally commissioned external review found, no clear terms of reference. The gender liaisons were marginalised from Center decision-making, and they had other responsibilities and other bosses to deal with, and so had little chance to influence the gender agenda.

Although the climate for gender research in the CGIAR has considerably improved in the past five or six years, especially in terms of funding, young gender researchers still face problems. GENNOVATE is a global research project aimed at enabling equality in agricultural and environmental innovation through an understanding of gender norms and agency, which began in 2013. It involves a majority of CGIAR Centers and Research Programs and many of the researchers involved in the project are fairly junior. Whereas in some cases these young researchers have received generous financial and professional support from CRP leaders, in other cases they have had big struggles with securing resources and have not been able to make the types of time commitment which have been really needed. They've had to prioritise other work because they didn't have the voice to argue for the importance of their work for GENNOVATE.

One of the factors behind this is the historical development of gender within the CGIAR. Except for a few cases, there have not really been career paths available for gender specialists which would lead to the existence of significant numbers of senior gender specialists within the system. I was responsible for recruiting the first gender specialists in CIP. The first direct hiring of an individual with a full gender specialist portfolio in CIP was in 2013, so it's a recent phenomenon. There are other Centers of the CGIAR with longer track records of hiring specialists to work on gender (CGIAR-IEA (2017),* but even in these Centers, senior staff like myself working on gender were hired originally as economists or cultural anthropologists rather than gender specialists. The 2010 annual portfolio report of the CGIAR system noted 'the need to develop proper capacity to address gender equity in research and product development' and that the individual Program reports 'provide fragmented snippets of gender-related research derived from milestones (activities) scattered in Program logframes' (CGIAR Consortium Office 2012: 8). But definitely, there was an important improvement in the situation with the establishment of the Consortium Gender Network in 2012, coordinated by Jacqueline Ashby. The network's initiative in establishing post-doctoral fellowships for gender specialists was an important contribution to bring in more gender specialists, albeit new junior positions.

I think it remains to be seen how their careers will develop in the system. And the extent to which, when there's funding pressure, gender positions will be retained. That's really an important issue. In CIP, my position has not been replaced because of funding constraints. So there's been a sort of erosion of people in senior positions with a strong gender focus, or with strong gender responsibility.

Babette: I don't think this is a situation exclusive to environment organisations, CIP or the CGIAR, Gordon. I think that in many places, gender focal points are people who are tasked to do gender mainstreaming on top of other duties

* This assessment of gender in the CGIAR was regrettably based only on the CGIAR Research Programs (CRPs) rather than the Centers where staff are actually hired, so this is an indirect view of the growth of gender capacity within the CGIAR System

> and where there are gender specialists, their jobs are the first to go when funding thins out. This is widespread even in many government organisations.

To conclude

We began by asking whether epistemic authority is masculine. Our discussion in this chapter shows that it is not masculine, in an essentialised, permanent sense. An intersectional analysis makes us understand that the social context and the specific dynamics of knowledge and power co-create and interlock multiple subjectivities that are not fixed but are changing and in constant relation with others. This analysis allows us to foreground the political and social dynamics that lead to experiences of marginalisation and privilege. Gordon's long career history highlighted the complexity and fluidity of professional identities and interactions in relation to the hierarchies of knowledge in his organisation.

Our conversations also discussed that breakthrough moments through mutual learning and joint efforts at research integration between gender and technical professionals are clearly possible, especially in an enabling policy environment. But, realistically, as Gordon's accounts indicated to us, there are limits for as long as epistemic inequalities – that translate to resource inequities – underlie the trajectory of agricultural research for development. Gordon sums up the possible reforms in this field:

> One of the conducive factors is the intensive interpersonal collaboration on the ground. Undoubtedly a second conducive factor is the control of funds in the hands of researchers who want to include gender in the analysis in the research work. If you look at the overall goals of the CGIAR and their alignment with the Sustainable Development Goals, the commitments to poverty alleviation, to equity, all those commitments clearly require gender to be a central part of research. But the problem is that many biophysical scientists do not yet make that connection, or if they make the connection in theory, it is still difficult to allocate funds for gender in the implementation of projects that they lead. This will require parallel changes at institutional and policy level that prioritise resource allocation, monitoring systems and personnel and project evaluation.

References

Biggs, S. (1990). A multiple source of innovation model of agricultural research and technology promotion. *World Development*, *18*(11), 1481–1499.

CGIAR–IEA (2017). *Evaluation of Gender in CGIAR*. Rome, Italy: Independent Evaluation Arrangement (IEA) of the CGIAR.

CGIAR Consortium Office (2012). *CRP Portfolio Annual Progress Report 2011*. https://cgspace.cgiar.org/bitstream/handle/10947/2759/CRP_Portfolio_Annual_Progress_Report_2011.pdf?sequence=4&isAllowed=y

Chambers, R. (1983). *Rural Development: Putting the Last First*. Longman.

Cho, S., Crenshaw, K. W., & McCall, L. (2013). Toward a field of Intersectionality Studies: Theory, applications, and praxis. *Signs: Journal of Women in Culture and Society*, *38*(4), 785–810. https://doi.org/10.1086/669608

Fraser, N. (2009). Feminism, capitalism and the cunning of history. *New Left Review*, *56*(2), 97–117.

Funk, N. (2012). Contra Fraser on feminism and neoliberalism. *Hypatia*, *28*(1), 179–196. https://doi.org/10.1111/j.1527-2001.2011.01259.x

Garry, A. (2011). Intersectionality, metaphors, and the multiplicity of gender. *Hypatia*, *26*(4), 826–850. https://doi.org/10.1111/j.1527-2001.2011.01194.x

Haraway, D. (1988). Situated knowledges: The science question in feminism and the privilege of partial perspective. *Feminist Studies*, *14*(3), 579–599.

Horton, D., Prain, G., & Gregory, P. (1989). High level investment returns for global sweet potato research and development. *CIP Circular*, *17*, 1–11.

Hovorka, A., Zeeuw, H. de, & Njenga, M. (Eds.) (2009). *Women Feeding Cities: Mainstreaming Gender in Urban Agriculture and Food Security*. Practical Action Pub.

Lykke, N. (2010). *Feminist Studies: A Guide to Intersectional Theory, Methodology and Writing*. Taylor & Francis. https://books.google.co.th/books?id=CkpMOsl32HoC

MacGregor, S. (2010). Gender and climate change: From impacts to discourses. *Journal of the Indian Ocean Region*, *6*(2), 223–238. https://doi.org/10.1080/19480881.2010.536669

Njenga, M., Karanja, N., Prain, G., Gathuru, K., & Lee-Smith, D. (2011). Community-based wastewater farming and its contribution to livelihoods of the urban poor: Case of Nairobi, Kenya. *Journal of Agriculture, Food Systems, and Community Development*, *1*(3), 151–162. https://doi.org/10.5304/jafscd.2011.013.004

Prügl, E. (2015). Neoliberalising feminism. *New Political Economy*, *20*(4), 614–631. https://doi.org/10.1080/13563467.2014.951614

Rathgeber, E. M. (1990). WID, WAD, GAD: Trends in research and practice. *The Journal of Developing Areas*, *24*(4), 489–502.

Sismondo, S. (2004). *An Introduction to Science and Technology Studies*. Blackwell Publishing.

Vita, L. D., Sciannamblo, M., & Viteritti, A. (2016). Re-thinking intersectionality through Science and Technology Studies: Trajectories of women in technoscientific fields. *Rassegna Italiana Di Sociologia*, *3*, 503–524. https://doi.org/10.1423/84375

3 Epistemic crossings of a marine biologist through gender encounters

By Bernadette P. Resurrección and Rebecca Elmhirst
In conversation with Maeve Nightingale

Becoming a gender expert

Gender mainstreaming today is almost an institutional pillar in development organisations, including in environmental conservation groups such as the International Union for the Conservation of Nature (IUCN). Conventional technical approaches in efforts to protect and conserve ecosystems and biodiversity have increasingly integrated social approaches, and gender mainstreaming ensures that gender issues are included. Apart from hiring dedicated gender specialists and professionals, environmental organisations sometimes assign their own technical personnel to conduct and carry out gender mainstreaming tasks. This has been the case with Maeve Nightingale at the IUCN Asia regional office who is a marine biologist by training, but who has become the regional office's gender focal point.

Some question whether a professional background is required to become a gender specialist. A useful way to tackle this question is how one acquires the knowledge to do this work effectively, which may not be the same for everyone. This chimes in with principles in feminist epistemology that posit knowledge as being situated and that there are multiple ways of knowing (Haraway, 1988; Trojer, 2014; Barbour, 2018). Additionally, this chapter will also explore how feminism and its ideas have been professionalised and like other professional sectors, its members may also seek ways to legitimate knowledge and expertise (Hoard, 2015; Kunz et al., 2019).

Maeve is senior and technical officer for coastal and marine programs at IUCN, which involves integrating ecosystem-based approaches with socio-ecological systems through a program implemented in 11 countries. She is also the regional gender focal point, integrating gender in all programs and building institutional capacity for gender analysis in the regional office as she works with senior program officers.

Maeve's particular route to becoming a gender expert was through field exposure and personal engagement with rural coastal communities for 22 years. It was not through formal training such as through the social sciences or gender studies. This aligns with findings from Bustelo, et al. (2016) that there is no agreed definition of the term 'gender expert', with no common criteria of what their work entails, and with no defined boundaries from where claims to intellectual authority may come.

Maeve: These last five years, I have devoted time to strengthening the resilience of coastal dependent communities in Asia. My role has been to integrate gender in the 11 countries where IUCN's Mangrove for the Future (MFF) operates, and as part of IUCN's gender and development policy. I have no formal background on gender, but only use my own observational skills having worked with coastal communities for 22 years, with the lens of women and men's roles and the balancing of societal relationships. So, when our regional director gave me the mandate to be the regional gender focal point for IUCN in Asia, I embarked on a three-year campaign to educate myself. Which meant that I talked to gender experts and mentors, taking their advice and reading a lot, so I could at least understand the language and adapt it to my own work and context.

I always had an interest in gender while working with coastal communities. I lived for eight years in these communities. Being a woman myself, the entry point was to sit and talk to women in particular. Without having any framework of analysis in mind, I was observing where the gaps were in terms of opportunities for women to have their voices expressed more publicly. And without knowing that that's called a 'gender gap', I was listening to women's stories juxtaposed with men's stories. In the fisheries sector, the teams I was part of were talking to the men. And I was talking to the women. And the fact that these two sets of exposure were so disjointed when they translated into development planning was quite striking to me. While working in this program, the regional director gave me the specific challenge to put things together in a sensible way and learn the 'gender' language. Honestly speaking, gender to me was an untouchable thing because I had no sufficient confidence to use the right terminology. This was quite daunting. It was like you had the experience and knowing what needs to be done but you didn't have the language! Until you had the language and the analytical framework to be able to express your observations, it felt that I was just on the doorstep of this process. I'm still not there yet, but I think I'm getting there.

Becky: So, is your professional identity that of a gender expert or a marine scientist? Are you becoming a gender expert?

Maeve: That will depend on who you ask that question to. Different people in my organisation will see me as the gender person, like the water program for example. But I see myself as a marine specialist. My confidence lies with marine resources management. But then having got this far in the gender integration work and being able to unpack gender in what we do, and to be able to bridge those two worlds, I would never close the door to being referred to as a gender professional. Since once you've opened it, you've opened it. And I think you will never walk backward. So, I wouldn't ever see myself dropping it since I really like it, and it allows me to raise the profile and concerns about marginalised groups in general.

Becky: When were the instances that you've found it not very helpful to identify yourself as somebody working on gender?

Maeve: I don't know if I do self-identify as a gender person in the first place [laughter]. That might be why I can't answer the question. But I do notice that within our own organisation as people kind of get to know that we're doing more and getting more practically experienced I'm now becoming known as the gender person. And sometimes, I feel like, 'Okay. That's nice and interesting.' But I don't want to subsume it under where I have more confidence in, which is in being a marine biologist. I don't mind that I've sort of become recognised as the gender person and not the marine person [laughter]. I think, in general, I don't think it hurts. I wonder sometimes though – I try not to forget what it was like not understanding anything about the language that goes with gender and development. I also don't think that I would stand up in an international conference and profess to be a gender person. I would explain my ideas on gender only on the basis of my own experiences.

Babette: But now you have a new lens. We gender folks working in environmental and natural resource management contexts usually need to learn about biophysical and technical conditions because gender and social issues are embedded in them. But that doesn't make us technical specialists, I'm afraid.

Maeve: And the lens is becoming more comfortable. But I know how it felt talking to gender people at the outset. And sometimes, it can feel that there is a barrier. Well, not a barrier, maybe a kind of hierarchy. There is a kind of supreme idea that either you are, or you are not, a gender expert.

Babette: Really? I thought the opposite [laughter]. I thought that it was we gender experts who are the ones most often edged out.

Maeve: I don't think it [the hierarchy] would be something that was easily confessed. But I think gender people who only know gender, who don't necessarily have an interdisciplinary background, their passion, their commitment is interesting and it's all cool. But there's also a bit of … (voice trails off).

Babette: Arrogance?

Maeve: A little bit.

Babette: Maybe it's also because of our defensiveness? We need to protect the space? Years and years of being called 'flaky' by the sciences have probably made us defensive (laughter).

Maeve: It's a patch. It's a turf.

Babette: Oh, so it's territory.

Maeve: Yes, maybe it's that. The majority [of gender professionals] though want bridges to be built. But I remember an incident when we were producing training materials that link gender with the UN Convention on Biodiversity. We were all in a room for three or four days putting forward ideas and materials. On Day 4, something struck me: there's not a word for 'gender' in many languages. And I thought that was kind of interesting. There are sets of words, substitutes and also descriptions. But that wasn't really the point. So,

I remember on Day 4 reiterating this comment. I said, 'If there's not a word for gender, shouldn't we unpack why there isn't a word for gender?' Meaning isn't there a reason why people do or don't understand gender simply by the fact there's not a word for it? Could that help us to grow an avenue of capacity development if we know why?

And there was a woman there who had worked for a major organisation, a very nice person, very knowledgeable, 'very gender', (laughter) and she said, 'We have fought long and hard to get gender on the table. It's in the policy dialogue. It doesn't matter that there's not a word for gender. This is not the issue. Gender is a fact. It is an established concept.' I was dumbfounded by her remark. I tried to reinterpret what I said that might have elicited that response. I was quite embarrassed, and I said, 'Oh, I'm sorry. I didn't mean to offend you.' I was just thinking about how we can deal with people not really knowing about gender since there's no word for it in some languages.

I was left completely in the dark. And even until today I feel it was a really intimidating experience. And I think that just reflects my own lack of confidence. It's probably not how gender people mean to project themselves. It's just my own lack of confidence in that environment. And none of the gender people batted an eyelid as to why this lady was kind of peeved. And I couldn't really get anybody to explain it to me. And I didn't want to belabour the whole thing, so I just let it go. It was fine. It was okay.

Babette: No, no, no. For me, that's not fine. It's silencing someone.

Maeve: In that workshop, I felt that I've had such limited experience. I think if you hang around gender people all the time, it becomes a language and a way of thinking, patterns of thinking that are common and comfortable. But if you impose it on environment language and ideas, the question arises on how we make the links. How do you put in gender into environment? I'm coming from an environment perspective, trying to find the entry points where it fits most comfortably and make the quickest sense to people.

This interesting episode may have indeed unleashed the defensiveness of gender experts towards questioning the very object of their work: that is, veracity of the existence of gender. As queer theorists and some feminists question the stability of 'gender', and the power dynamics that underlie its materialisation (Butler, 2004), sometimes through the very application of 'gender mainstreaming', gender experts may rightly need to reflect on their work and practice (Zalewski, 2010; Davids et al., 2014). Destabilising gender prompts a range of responses, and this gender expert may have experienced being pushed to the margins in the past and was not going to allow it to happen in that meeting. However, this explanation is insufficient and begs that we explore other elucidative avenues. One promising avenue is to explore how certain narratives and knowledge become 'expert truth', including, and not exempting, gender expertise.

Scholars who study professions have often observed how experts or professionals create boundaries to secure monopoly over their expertise and to protect its autonomy and authority. Claims to expertise empower professional groups to draw and defend boundaries that set conditions for inclusion and exclusion (Azocar & Ferree, 2016; Liebrand, 2014). That said, other discursive elements also came into play in the episode just narrated and which served to silence. First, by invoking that gender is a 'fact', the speaker was mimicking the 'God trick' (Haraway, 1988), using the positivist argument where veracity, truth and authority are only based on what is observable, objective and measurable, or in short, it was an attempt to vest gender knowledge with the power of scientific truth. Ironically, feminist epistemology has long argued that knowledge is situated and not value-free. Second, science, technology and society (STS) scholars contend that expertise often operates as a network of power (Brady, 2018). In this view, stating 'gender as a fact' summons the historical assemblages of artefacts behind this statement such as gender indicators, gender analyses tools and checklists, academic and non-academic training curricula, publications, gender experts and feminist movements, donors, institutions all of which have been enrolled in the making of this networked field (van Eerdewijk & Davids, 2014; Cornwall et al., 2007). Lave (2011, 2012) presents a similar case through her analysis of how a specific yet heavily contested type of geomorphological expertise on stream restoration came to be dominant. This expertise was legitimated through an assemblage of non-degree courses, publicity stunts, accreditation procedures, and social capital among trainees who later became well placed in US federal agencies. Gender expertise can also build from the accumulated performance of an actor network and its own assemblage of artefacts (e.g., mainstreaming tools, gender concepts and so on). That said, it is worth pointing out that gender expertise is also often unstable and comes into view when it is 'produced' relationally, through power-laden webs of political and social relationships which, for example, also defined the 'gender-is-a-fact' episode.

Maeve also expresses uneasiness with efforts to merge the social with the technical, and thus faces the difficulties of 'translating' the social or gender knowledge in an environmental action context.

Babette: We're coming from the other side. For instance, we try to look for how can gender equality be realised in natural resource management contexts. Let's say in agriculture: issues of labour, rights to land, may be the entry points.

Maeve: But those points are still sometimes a separate matter. They all make sense by themselves. But fitting them in, finding the flow between environment and gender issues, it feels a bit of a 'hit or miss' sometimes. And I find the same thing in gender trainings – most trainers take you into a framework, but that doesn't leave anybody with knowing what to do. Social and gender issues have a separate dialogue from environmental issues. They make sense on their own. How do we integrate these with the scientific aspects of the environment? Where is the commonality? And must there be?

Now I'm thinking of our gender equality tool at IUCN and how it actually doesn't agree with coastal issues from the scientific end. The tool is still predominantly social.

We learn through trial and error which bits would work and which bits are more powerful, and how to facilitate those which are going to give you the quickest insight and exchange among people.

In a related chapter in this book, Margreet Zwarteveen and Seema Kulkarni discuss how it may be fruitful to recognise the plurality of knowledges rather than attempting to integrate them. It means creating room for different ways of knowing and different versions of reality to co-exist. Knowledge acquired through experience – or embodied knowledge – demonstrates the plurality of knowledges.

Embodied knowledge

From her experiences working with coastal communities, Maeve has observed mismatches between program goals and gender realities. She encounters migration, the loss of land, personal insecurity, and gender-based violence that are unfortunately beyond the purview of her organisation's conservation goals and livelihood programs. There is also the issue of short project durations that may be unable to effect substantial change and Maeve is acutely aware that transformative changes take time. She nevertheless takes note of the small changes in these communities she spends time in, which in the long term, she believes may be incremental.

Maeve: I have witnessed small changes along the way: women become more confident to speak and represent themselves. They're also able to talk about domestic violence. They also try to find solutions jointly for issues that affect them, whereas otherwise they would have to suffer individually. It might be hard to imagine but on a three-year program basis, I think you could start seeing change, and certainly better representation of women outside their own households within a small part of the village. The most striking would be how readily women are able to express issues that they're experiencing, secretly, in confidence sharing with other women with whom they may never have talked about these things before.

Field experiences and close interactions with women can open new avenues of knowledge. Conventional scientific epistemologies cannot account for women's experiences and sometimes do not consider them as legitimate knowledge, as scientific knowledge is traditionally premised on objectivity and rationality. The dualistic ontology of Western knowledge traced back to the ancient philosophical works of Aristotle and Plato recognises reason and experience as being diametrically opposed. In turn, the feminist critique to science has long argued for the reconstruction of knowledge itself, alluding to other equally legitimate forms of knowledge, and more recently attempts to pluralise knowledge (Mukhtarov & Gerlak, 2014).

The feminist adage 'the personal is political', puts forth the idea that personal experiences of women are a source of important knowledge, and that they can be understood socio-culturally and are politically contextual largely because they unpack situations of inequality. 'Women's problems' – such as domestic violence – have come to be seen as widely structural, and therefore political (Barbour, 2018), thus meriting institutional attention and redress. Knowledge embodied in experience, feminists argue, is a legitimate form of knowledge. Gender experts in technical environment organisations who recognise the importance and relevance of this knowledge may often face the prospect of their findings not being given sufficient importance.

Babette: How then do you integrate gender with your coastal management projects on the ground and with your colleagues?

Maeve: I think that you shouldn't start out jumping off with 'gender' with people since they'll immediately think 'women'. It is better to start by talking about and understanding empowerment and disempowerment and what causes them among different groups who are commonly marginalised and experiencing disempowerment and the roots of that. Then one should go to how their marginalisation is being affected by their being male or female. So it's like looking at that minority group angle.

With colleagues, I feel that if you take them (our whole organisation is made up natural scientists), to 'let's look at how you can integrate a gender perspective', they'll only get as far as pointing out the women's roles, and perhaps the possibility of doing some women's projects. By doing this, the full analysis of gender inequality will be lost. For example, we saw in Laos that there is legal recognition for equal institutional positions for women and men, yet we see that at the community level there are predominantly male leaders. Why is that? So unpacking the causes of that is helpful, which helps us to think: what can our little project do in three years that might lead to an incremental stage towards enabling the leadership capacity of women and leveraging opportunities for women to have a stronger voice? So they think – *ahh* that means that when we have our consultations, we can address that gap. Then they think – but women can't talk, they have caring tasks and can't come – so what can our project do to address that? So after one day of this they went away feeling empowered and looking at their project in a different light, which was not prescriptive, and they felt capacitated to take on-board the thinking about inclusive development. And that's the key, right? It's being development practitioners and thinking holistically.

Babette: Considering that you've had 22 years of living with coastal communities, how much does planning really lead to change? As we know in real life, changes are usually unplanned, and life is much more complex, and changes actually come in unintended ways. So, will planning change really matter in the long run, you think?

Maeve: I think if we're talking at a project scale of three to five years, with some concrete objectives, and positioned in the context of an overall theory of change, I think planning definitely has a place, it can influence change. And if we're talking about specifically gender integration in some of the natural resource management ecosystems-focused projects, without planning concretely for social engagement, social empowerment, it won't happen by magic. So 'not' planning has an effect. You will get perhaps fewer rich results, definitely less sustainable results. Planning for short-term changes within a project context, and within the larger, longer term theory of change that you're working with in that context definitely can improve results or create positive change for people. Thinking about the gender context again, if we apply gender inclusion in our own processes for implementing projects, in project planning and design, already we will provide completely different outcomes. In communities, there are a few avenues for creating strong, empowered relationships.

Becky: Can you tell us about some examples?

Maeve: We went to an island called Nijhum Dwip, a southern island in the Bay of Bengal in Bangladesh. It's the most remote island you can get to from Dhaka and it's known as the quiet island. It was only actually created in the last 50 years and government programs resettled displaced people on this island. So, this community hasn't evolved into a socially cohesive community. People were resettled there generally because they have not been able to stay on their original homeland and many of them have weak land tenure or security. On top of this, women have very, very restricted mobility and this is having a direct impact on their opportunities to make change at the household and community levels, almost a complete invisibility of women's voice in societal decisions.

All of the gender tools that people talked to me about in the past suddenly made perfect sense. First of all, just create dialogues around basic human rights and build confidence among women facing the same issues within their homes who are not able to voice them, talking about the conditions that they live in and slowly coming to realize that those are not conditions that they necessarily need to accept or be expected to accept. We have a five-year project in the island that focuses on fisheries and governance of marine protected areas, but without us taking a really considered effort to have long-term 15-year commitment to work with those communities, we're not going to see much change.

Becky: We've heard this conversation with others around how it is sometimes almost better to engage with these communities with a sort of broader view on social transformation rather than just using the gender as your entry point. And your example kind of gives suggestions to why the gender word is necessary in a very clear way. But I'm guessing that in other places where you've worked that perhaps it isn't quite so obvious. So, in terms of planning for changes are you also not focusing on bigger forms of empowerment other than gender equality? What's your thinking on that?

Maeve: These women on the island are severely constrained in their ability to broaden their opportunities. Once they start to actually increase their latitude, they are set apart socially and stigmatised. And in that sense, it becomes much more counterproductive to plan for change given those conditions, but other entry points might be possible. I guess that's what change being so complex means. So, we aim to create some sort of safe space. What these women are experiencing – domestic violence – is just laughed off and continues to be passed down by generation and completely explainable by these ladies. And it's not just one or two, it's across everybody we interviewed. When at 13 if you're not married then you become a burden. Women speak about their own daughters being burdens. They talk to us about these issues despite being strangers. They would say, 'You can understand why we get hit – when we haven't got food on the table and there's no petty cash for cigarettes, you can understand why they're annoyed.' They also said that 'the water level has raised, and we would need to lift our saris to get to the tube well and we can't do that. It's rainy season. If we keep our saris down, we are then going to have to sit in the one and only sari we have which will be completely wet.' So, they can't fetch water. If they don't fetch water, nobody at home has water, right? They discuss these dilemmas as a group and there would be solutions that they find. These are things that you can find solutions for if you work together. But if you're on your own suffering individually it's hard to imagine what on earth is going to change that scenario.

Becky: How do you reconcile your technical background and your community work?

Maeve: Okay, you're referring to being a natural scientist turning into a social scientist, right? In teaching integrated coastal management at the Asian Institute of Technology, for example, we teach all the natural science part such as the ecosystem structuring function. Then we also teach the social science part of it which is understanding the communities you're working with and providing them with principles and tools to do that, which most of the foresters and the fisheries graduates almost have never come across. Science is maybe perceived to be technically complicated, and social science is perceived as the soft science, so to speak. But to me, what's more important is where the meeting place is. And in our organisation, we deal a lot with governance such as the protection of biodiversity through good governance. And in unpacking the governance part of it we come to understand social equality issues.

I remember in one field situation, our project manager requested that we de-brief him on what we found out from our gender study. After listening to our accounts, he said, 'Okay, that's nice. So, you've turned up the fact that there's domestic violence issues, and you've turned up the fact that women have limited access to education, and that child marriage is a problem, and mobility. But we don't want to get too waylaid by that. We want to focus on the governance of marine protected area.' When he said that I and my companions were aghast. I think he missed the point. We need to understand how communities can change their circumstances, right? All the rest might follow.

Maeve's stories of her immersion in coastal communities reveals the workings and importance of embodied knowledge. To effectively engage in pluralising knowledge, Mukhtarov and Gerlak (2014) say that policy professionals should be able to tell compelling stories rather than relying on the 'bloodless discourse of science and expertise'. Her encounters with local people in the coastal areas offer her 'practical knowledge' derived from her first-hand observation of the social contexts within which coastal people use their resources. 'Knowing-through-practice implies that knowledge is specific to a particular situation and, therefore, tied to specific communities. In this way, practice is not intended to reflect "best practices" or more cookbook approaches. Rather, it is through interaction that people make sense of reality, negotiating the meaning they give to their surroundings and their actions.' (Brugnach & Ingram, 2011: 52).

To conclude

We learn from Maeve that the route to the gender profession can be through one's keen interest and immersion in the daily lives of people in communities. Immersions in communities or 'life on the ground' leads to insights and reflections that can enlighten action and practice.

However, acquiring new knowledge, such as gender knowledge, can be challenged by its own vanguards as its professionalisation like in all other sciences evolves its own forms of exclusions and inclusions.

From the examples presented in this chapter, we have seen how professionals may contribute to the building of silos between and within their sciences through networks and other artefacts. There is need to respect the different sciences, their different and multiple ways of knowing, since erecting professional walls serves only to exclude and slow down the success of development in more holistic and inclusive terms.

References

Azocar, M. J., & Ferree, M. M. (2016). Engendering the sociology of expertise. *Sociology Compass*, *10*(12), 1079–1089. https://doi.org/10.1111/soc4.12438

Barbour, K. (2018). Embodied ways of knowing: Revisiting feminist epistemology. In L. Mansfield, J. Caudwell, B. Wheaton, & B. Watson (Eds.), *The Palgrave Handbook of Feminism and Sport, Leisure and Physical Education* (pp. 209–226). Palgrave Macmillan UK. https://doi.org/10.1057/978-1-137-53318-0_14

Brady, J. (2018). Toward a critical, feminist sociology of expertise. *Journal of Professions and Organization*, *5*(2), 123–138. https://doi.org/10.1093/jpo/joy004

Brugnach, M., & Ingram, H. (2011). Rethinking the role of humans in water management: Toward a new model of decision-making. In *Water, Cultural Diversity, and Global Environmental Change* (pp. 49–64). Springer Netherlands.

Bustelo, M., Ferguson, L., & Forest, M. (Eds.) (2016). *The Politics of Feminist Knowledge Transfer: Gender Training and Gender Expertise*. Palgrave Macmillan.

Butler, J. (2004). *Undoing Gender*. Routledge.

Cornwall, A., Harrison, E., & Whitehead, A. (2007). Gender myths and feminist fables: The struggle for interpretive power in gender and development. *Development and Change*, *38*(1), 1–20. https://doi.org/10.1111/j.1467-7660.2007.00400.x

Davids, T., van Driel, F., & Parren, F. (2014). Feminist change revisited: Gender mainstreaming as slow revolution. *Journal of International Development*, *26*(3), 396–408. https://doi.org/10.1002/jid.2945

Haraway, D. (1988). Situated knowledges: The science question in feminism and the privilege of partial perspective. *Feminist Studies*, *14*(3), 579–599.

Hoard, S. (2015). *Gender Expertise in Public Policy towards a Theory of Policy Success*. Palgrave Macmillan.

Kunz, R., Prügl, E., & Thompson, H. (2019). Gender expertise in global governance: Contesting the boundaries of a field. *European Journal of Politics and Gender*, *2*(1), 23–40. https://doi.org/10.1332/251510819X15471289106112

Lave, R. (2011). Circulating knowledge, constructing expertise. In M. Goldman, P. Nadasdy, & M. Turner (Eds.), *Knowing Nature: Conversations at the Intersection of Political Ecology and Science Studies* (pp. 363–279). University of Chicago Press.

Lave, R. (2012). Bridging political ecology and STS: A field analysis of the Rosgen Wars. *Annals of the Association of American Geographers*, *102*(2), 366–382. https://doi.org/10.1080/00045608.2011.641884

Liebrand, J. (2014). *Masculinities among Irrigation Engineers and Water Professionals in Nepal* (Wageningen UR Library) [Wageningen University]. http://edepot.wur.nl/321002

Mukhtarov, F., & Gerlak, A. K. (2014). Epistemic forms of integrated water resources management: Towards knowledge versatility. *Policy Sciences*, *47*(2), 101–120. https://doi.org/10.1007/s11077-013-9193-y

Trojer, L. (2014). Gender research as knowledge resource in technology and engineering. In W. Ernst & I. Horwath (Eds.), *Gender in Science and Technology: Interdisciplinary Approaches* (p. 262). Transcript-Verl.

van Eerdewijk, A., & Davids, T. (2014). Escaping the Mythical Beast: Gender mainstreaming reconceptualised. *Journal of International Development*, *26*(3), 303–316. https://doi.org/10.1002/jid.2947

Zalewski, M. (2010). 'I don't even know what gender is': A discussion of the connections between gender, gender mainstreaming and feminist theory. *Review of International Studies*, *36*(1), 3–27.

4 Beyond the business case for gender

A feminist political ecologist in the FAO*

By Rebecca Elmhirst and Bernadette P. Resurrección
In conversation with Clara Mi Young Park

Introduction

A recurring thread that weaves through critiques of gender mainstreaming in international development organisations has been the charge that those tasked with 'doing gender' have been complicit in its bureaucratisation, thus inadvertently stripping out the possibility of transformative change (Cornwall et al., 2007; Fraser, 2009). Within environment and development contexts, the suggestion that the exercise of gender expertise may have turned feminist strategies from a model of resistance to an instrument of power is contrasted with the transformative environmental feminism explored in academic and activist feminist political ecology (Çağlar et al., 2013: 6; Arora-Jonnson, 2014; Collins, 2018). Effectively, the claim is that gender experts have been vulnerable to co-optation by privileging certain kinds of gender equality within their work, of the sort that effectively supports neoliberal capitalism, whilst constraining more progressive agendas. In this reading, so-called 'femocrats' have sold out the women's movement and are profiting from women's disadvantage (Chappell, 2002; Miller & Razavi, 1998; Cornwall et al., 2007). However, others have suggested that this vision of gender expertise is too narrow. Whilst a simple binary distinction between the femocrat and the feminist may hold when observed from a clear 'outsider' academic position (Ferguson, 2019), the picture painted in the personal accounts of feminist gender experts is more complex, pointing instead to the fine details of compromise and negotiation that make up gender mainstreaming and equality practice (Ferguson, 2019; Eyben & Turquet, 2013). The figure of the femocrat or bureaucratic mandarin that has been the focus of critique dissolves as gender advisors move between feminist and bureaucratic subject positions across time and in different spaces, in ways that perhaps accord more closely with Sandler's argument that femocrats are 'the warriors within' (Sandler, 2015).

In this chapter, we reflect alongside Clara Mi Young Park who, at the time of our discussion, was working as regional gender officer in the Asia-Pacific office of the UN Food and Agriculture Organization (FAO). Alongside her terms of reference and responsibilities in this role, she was also writing her doctorate in

* The views expressed in this chapter are those of the author and do not necessarily reflect the views or policies of the Food and Agriculture Organization of the United Nations."

Feminist Political Ecology, based on research on the gendered impacts of large-scale land deals. Combining these related but very different aspects of gender work have led her to observe that she was 'wearing a double hat'. Our discussions focus on the work Clara does around supporting gender equality within the specific context of the FAO, an institutional environment that on the one hand is dominated by technical fields and engagement with sometimes conservative country ministries, but on the other, is part of the UN, and thus influenced by the latter's function in developing and steering the international governance of gender, sustainable development and human rights (Sandler and Rao, 2012; UN Women, 2018). As Ferguson (2019: 383) argues, the application of gender expertise in complex multilateral contexts (such as the FAO) and on the ground may involve negotiating definitions of what is 'okay' and 'not okay' to say about gender (Eyben, 2010: 65), or selling gender through 'tactical slogans', which has included making the business case for gender (Chant, 2012: 201). In our discussion, we ask Clara to reflect on the steps she is making to advance gender equality within the FAO. Amidst inevitable bureaucratic compromises that arise when working in a complex multilateral and multidisciplinary organisation based in technical environmental science disciplines, are there breakthroughs from which more transformative change can be leveraged? How do these sit alongside the engagements she is making with academic feminist political ecology through her PhD research and publications? In other words, what advantage can be achieved from wearing her 'two hats' as she sets her sights beyond the business case for gender in the FAO?

The chapter is organised around three themes that emerged when we explored the latitude and openings Clara has been able to identify for inching towards transformative change, within the context of FAO policy on gender, and through the constraints and opportunities of working with key decision makers within national governments. The first of these is her approach to working alongside the 'business case for gender' that threads through the FAO's gender policy. In the section that follows, she shares her perspective on the difference the Sustainable Development Goals (SDGs) have made for pressing this case in new directions and the structural changes that are needed to enable this on the ground. Finally, Clara shares her thoughts on the soft skills that are necessary for effective feminist work within and through an organisation like FAO which involves collaboration with colleagues and partners coming from technical science disciplines. We begin by asking Clara to explain her work with FAO as a researcher with a background in feminist political ecology and women's rights, and what this means in terms of wearing her 'two hats'.

Working as a regional gender advisor in the FAO

As a specialised agency of the United Nations, the Food and Agriculture Organization (FAO) has placed gender equality centrally within its mandate to achieve global food security for all by raising levels of nutrition, improving agricultural productivity and natural resource management, and improving the lives of rural populations (FAO, 2018). In its own words, the goal of empowering women and girls underpins its approach to reducing gender inequalities, through its work on norms and standards, data and information, capacity development, knowledge and technologies, partnerships, advocacy and communication. As a specialised agency of the United Nations, the incorporation of gender within FAO's mandate shares a history with the UN more broadly. Skard (2009) notes that although the original 1945 UN charter made reference to women's rights, this was not the case for the FAO, and no mention was made at its first conference in 1949. Since then, however, the influence of the UN's women's rights agenda has been felt within the FAO, particularly through the UN Decade for Women (1976–1980), the intergovernmental Commission on the Status of Women (CSW) and the UN Convention on the Elimination of All Forms of Discrimination Against Women (CEDAW). Change had gathered pace thanks to landmark studies (e.g. Boserup, 1970), and through the UN's global women's conferences through the 1980s and 1990s, as the FAO itself shifted from gender-blind development to approaches emphasising women's inclusion (women in development), the structural issues underpinning gender equality (gender and development) and empowerment (Skard, 2009). The Beijing Platform for Action, agreed multilaterally at the UN's Fourth World Conference on Women in 1995, galvanised gender equality as one of the FAO's priority areas for interdisciplinary action, creating bureaucratic spaces in which to advance the gender agenda by mainstreaming gender equality within the organisation and by bringing gender equality into its technical work.

The main context for our discussions in this chapter has been the implementation of the FAO's (2013) *Policy on Gender Equality*, which followed its publication, the *State of Food and Agriculture 2010–2011* on women in agriculture (FAO 2011). The policy on gender equality sets out a framework for guiding FAO efforts to achieve gender equality, with the objective of closing the gender gap, including a commitment that 30% of total agricultural aid would be committed to projects related to women and gender equality. The FAO describes its gender policy as prioritising 'the equal participation and decision-making of women and men in rural institutions and in shaping laws, policies and programmes; the equal access to and control over productive resources, services, income, markets and decent employment; and the reduction of women's work burden' (FAO, 2018: 4), and sets out accountability measures for achieving this. At the time of our discussions, this policy was being updated to align with the SDGs to strengthen efforts at accountability.

The mechanism for implementing FAO's gender policy is through its Gender Network system, a network of officers and Gender Focal Points at headquarters and decentralised offices who devote their work or a percentage of their time (about 20%) to promoting gender equality in their technical areas of work. At the time of our discussion, Clara was working in the Asia Pacific Regional Office, as regional gender officer. With a background in gender and human rights advocacy working in non-governmental organisations, Clara began her work with the FAO as a consultant, specifically on the gender equality implications of large-scale land investments at the time when rising food prices had created a spike in so-called 'land grabs'. When the FAO's gender strategy was rolled out and the gender positions in the regions were reinstated, she was hired as regional gender officer. We began our discussion by inviting Clara to describe what her work involves.

Clara: I'm here as the regional focal point for gender, following the audit of gender work that FAO did between 2011 and 2012. When the corporate policy on gender equality was published, one of the recommendations picked up was that there should be gender advisors in all the regional offices. Some of the positions were already there, but vacant or not filled, some positions were not there at all. This also was part of a broader effort to strengthen gender work: the institutional capacity and the gender mechanism within FAO.

In parallel with FAO headquarters, Clara's work is divided into two main areas, first relating to mainstreaming and the institutionalisation of gender within the FAO's work in the Asia Pacific region and secondly gender work – research, advocacy and policy – within specific thematic areas.

Clara: We have a network on gender focal points in all the country offices, this was one of the recommendations of the FAO's gender policy that we have a network of people that could support at country level. These people are not doing only gender work but have other responsibilities. The change [brought by the new policy] was that there were Terms of Reference formulated for people appointed as focal points, and that these be professional staff at the P4 level or above, so not just the most junior staff or the secretary of the boss but people who were engaging in programming. In practice, this is not always the case, and not always possible, but at least on paper the intention was there. Gender work is also monitored in that they are accountable for giving 20% of time for gender. My work is also about coordinating with them, facilitating dialogue amongst them and to support them.

This type of coordination brings challenges with it, as Clara explains.

Clara: The support could be anything and everything. It could be related to a project, it could be on specific requirements that all country officers are supposed to meet as part of the gender policy, for instance. The network, as you can imagine, is made of a diversity of people and interests, expertise and backgrounds. There are people who are more committed and interested, others not as active or proactive. I'm also doing the same thing in the regional office, which means that we have teams working on natural resources and climate change, and on the agricultural side, nutrition and subsistence, and I'm supposed to do my mainstreaming work with them. I have to say that in a way I find it more challenging to work with my own colleagues in Bangkok than I do in the countries: there is some expectation that because I'm here, gender work is something that I do, not what they should do! There are of course exceptions to that. As part of this function there are also monitoring and reporting tasks which are partly linked to the implementation of the [FAO gender] policy, we also have a regional strategy and action plan which I'm also monitoring as well as other institutional requirements such as to review all the projects that go through the project review committee here at the regional level.

Alongside this mainstreaming activity, the second part of Clara's work is undertaken in specific thematic areas, including some normative work, publications, research collaborations, and training in specific technical areas, for example, value chains, climate change or forestry. Projects that she undertakes are categorised in two different groups, those focusing on gender or women's empowerment, and projects that are within any of the areas that fall within the FAO's mandate, that have a gender component or strategy. Some of the work in this second area is triggered by emerging issues, regional priorities, or triggered by opportunities, including collaborations with others such as research institutes or UN agencies, as the FAO is part of the UN-wide coordination mechanism working on gender equality. Clara notes the distinction between these two areas of her work is somewhat hierarchical.

Clara: There is a division even in HQ between people who work on mainstreaming or 'more technical work'. There is a subtle kind of discrimination as mainstreaming is seen as not technical enough, unlike those people working on land, fisheries, value chains, the specific sectors. Here, I do both!

As a regional gender advisor with a coordinating role, much of Clara's work involves her engagement at other levels, including country and international level. At an international level, her work is embedded in the implementation of the FAO Gender Policy which feeds into the UN system wide action plan (SWAP) for gender mainstreaming. This means she reports to colleagues in headquarters, providing information tracking how many trainings or country gender assessments have been completed, for example. At a regional level, she is part of the project review committee, where projects uploaded onto the FAO system are reviewed before implementation starts. Her work also involves taking stock of gender work, including through a web-based monitoring system of gender markers that is used to track, monitor and report on the extent to which a project addresses gender equality, ranging from this being a main focus to gender equality not being addressed. At country level, she works with UN country-level gender advisors and others, and their counterparts within national government departments and ministries.

The use of instruments and tools to mainstream gender described above feels distant from the academic literatures and practices of academic feminist political ecology that Clara has deployed in her research outside the FAO. Her personal research has centred on gender and generational dynamics in relation to large-scale land deals in Southeast Asia and beyond, working in collaboration with activists and advocacy groups (e.g. Park et al., 2015; Park, 2019). We were interested in exploring the conundrums raised when a feminist political ecology activist academic works within mainstream environment and development work, given this is a path taken by many of the new crop of gender experts (Arora-Jonsson and Sijapati Basnett, 2017). Partly to answer this conundrum, Clara has adopted a strategy of separating her academic-activist world and her professional work at FAO, to give herself freedom in her academic research. The separation also reflects the difficulty of reconciling the different logics of mainstream gender programming in the FAO and academic-activist research, where the latter has depended on building long-term relationships of trust with communities impacted by and contesting large-scale land deals. Tacking between her largely ethnographic and critical discourse-related feminist political ecology work and her role as a feminist in a bureaucracy has meant confronting and embodying several points of tension that mark out the challenges of being a 'warrior within' (Sandler, 2015). However, what emerges in our discussion is how these tensions also crack open small fissures in which opportunities for small transformative gains may be found (Bergeron, 2016; Arora-Jonsson and Sijapati Basnett, 2017) and Clara's story has very much one of persevering towards feminist change. As she puts it, 'I don't want to give up on something that I'm doing just because I feel frustrated and we often feel frustrated doing this kind of work. So we also have to resist the temptation to just give up'.

Working with the business case for gender

An important theme that we explored in our discussions concerned the issues raised by the dominance of what has been termed the 'efficiency' approach to gender equality within gender, environment and development debates (Resurrección, 2017). Since 2010, the work of gender advisors at FAO has gained traction within the organisation and beyond, influenced by the view that gender equality makes smart economics (World Bank, 2006; World Bank, 2011). This was further endorsed by major development donors through the Busan Joint Action Plan on Gender during the Development Assistance Committee (DAC) meeting in 2011 (Eyben, 2015). As a form of neoliberalised feminism (Prügl, 2015), this approach has drawn much criticism from feminists (Chant and Sweetman, 2012; Çağlar et al., 2013; Parpart, 2014) and from within feminist political ecology (Harcourt and Nelson, 2015; Resurrección, 2017). Critics have focused on how so-called efficiency arguments for improving women's access to employment and to markets has dislodged an agenda of gender justice and ethics based on an analysis of structural and institutional drivers of gender inequality (Calkin, 2015). Instead, priority is given to the incorporation of women within existing neoliberal economic orthodoxies, with contrary and potentially damaging results (Cullen & Murphy, 2018). As Davids and van Eerdewijk put it, 'ironically, these win-win and efficiency gender narratives have opened up space for a celebration of women as change agents, either as gender experts in bureaucracies, NGOs, or grassroots organizations or as girls and women in their own families and communities. It is assumed that the disempowered will empower themselves, and meanwhile the power holders in the status quo are left off the hook' (2016: 89). They suggest that this interrupts the relationship between feminism and gender. This was likewise demonstrated by Cullen and Murphy (2018) through cases in the European Union pointing to the potential of the business case for gender equality to support the inclusion of women in governance structures, yet in the end these have led to perilous outcomes.

Despite these critiques, the business case for gender equality keeps being made over and again, perhaps with a less instrumentalist emphasis but centred around claims concerning the efficiencies afforded by tapping into women's and girls' potential. In our discussion, we were all aware that making a business case for gender equality softens otherwise stony ground for gender mainstreaming and the inclusion of gender analysis in technical programming. The challenges of bringing gender from the pages of corporate strategic plans and into everyday practice at FAO underpins why such discursive tools might be needed and how these become part of the bureaucratic compromise. Clara notes that, whilst work on gender equality is mainstreamed at the level of discourse, tools and knowledge products at FAO, the culture around doing gender has not completely changed.

Clara: I'm not sure to what extent people have really internalised [gender equality] and made it also their priority. I'm talking about colleagues and people working in the different technical disciplines [at FAO]. To what extent people have really bought into the importance of doing gender, and that it is not just the responsibility of a few people but is everybody's responsibility.

For feminists working within technical institutions where the gender agenda is a harder 'sell', this has presented a particular dilemma, and one where framing gender equality as efficiency takes hold. Not long after the World Bank made its business case for gender equality (World Bank, 2006), the FAO's flagship report *State of Food and Agriculture* (SoFA) (2010) was published, focusing on *Women in Agriculture*, with a framing that echoed the World Bank's 'efficiency' approach. This landmark publication has been identified by many in this sector, including other contributors to this book, as having helped change the landscape for gender-inclusive agricultural research for development, by providing leverage for integrating gender into their work. Was this a view that Clara shared?

Clara: To this day, some of the messages that came out of that report are the ones that we keep referring to. Really, because there was a business case for doing gender. From a feminist perspective, we don't like to say it is a business case. But it has really resonated with different people and different stakeholders. It was a message that was very easy to pick up and very easy to sell. From a feminist perspective, we want to really see the rise and the full empowerment of gender come to the fore. But then we're also stuck in a place where if you don't give these kinds of arguments – we keep seeing this repeated in other institutional reports, the World Economic Forum – if you don't keep this economic rationale, it seems like we are not able to push strongly, especially in specialised technical agencies.

Clara notes that this discourse has provided a very good way of making the case and selling the point of gender equality to partners, governments and donors, as well as internally within the FAO. The immediate reaction to this has been to use the tools and accountability mechanisms to ensure that the gender policy is adhered to. For example, one way was increasing the number of women beneficiaries in the context of ensuring that with more women in agriculture, productivity is made more efficient and increases the prospects of overall economic growth. Is Clara satisfied with this approach?

Clara: I think we are always going back to the same conundrum, between having achieved so much as in terms of institutionalising gender mainstreaming, having this package of tools, using these to make people accountable, which are useful in a way. But then not following through to take the extra step, to go beyond the mechanical collection of data and information. At the end of the day, I think especially in organisations like FAO, we are still at the point where we think we have made a lot of progress, and we have in many ways, but we are still not there in terms of seeing gender as a political issue, as something that needs to be addressed at different levels, rather than just the number of people attending a workshop or the number of beneficiaries. I recognise the value of what we have done, the usefulness of using these instruments for highlighting things we have done and what could be done in a different way and to raise a flag when things could be done better. I am certain that what we are doing is not enough.

So what would be enough? What additional steps is she able to take to move beyond the limitations of a business case for gender and the quantitative dimensions of gender mainstreaming, and towards transformative change? Although Clara notes that she does not have solutions, there are some key points that she identifies as having given her the space to shape the work she is doing more effectively.

Making a social sustainability case for gender equality

We were interested in hearing what other entry points Clara had found for making the case for gender equality in her work at FAO, and with the partners she works with at country level, and the ways these could help press the case being made in new directions. There has long been a recognition of the existential pain that feminists experience in technical and bureaucratic organisations where instrumentalist arguments are used to justify the inclusion of gender issues (Miller & Razavi, 1998; Eyben, 2010). Clara's exposure to realities on the ground as a field researcher in feminist political ecology deepens this concern. On the one hand, she has traced the adverse inroads of the capitalist economy on women's and men's lives in rural communities through land grabbing, whilst on the other, her role at FAO has involved her leaning on economic justifications such as the 'business case' for gender mainstreaming, prompting much reflection for her. Are there ways that she has been able to work around the downsides of the 'business case' in order to respond to some of the criticisms which, as a feminist political ecologist, she is acutely aware?

Clara: The Sustainable Development Goals have been helpful so I don't necessarily use the business case as much as I use the social sustainability case. It has provided a very good way of making the case and selling the points with governments and donors, as well as internally.

According to the FAO's documentation, which is key in norm-building, the principle of 'leaving no one behind' guides the 2030 Agenda for Sustainable Development and FAO's Policy on Gender Equality, which prioritises the equal participation and decision-making of women and men in rural institutions and in shaping laws, policies and programs; the equal access to and control over productive resources, services, income, markets and decent employment; and the reduction of women's work burden (FAO, 2018). This policy is currently being updated to align more closely with the SDGs. Why has this alignment been important for her in moving beyond the business case with its emphasis on women's economic empowerment? Clara suggests this is because the SDGs have enabled her to encourage partners to look behind the numbers of women beneficiaries, for example, and instead look at the quality of what is being delivered.

Clara: I would say a case in point is about creating more work for women without looking at what the constraints they face. Because these easy-to-sell points [on increasing numbers of women employed, for example] are not always well explored, the risk is that we end up doing something that is not really transformative of social norms and entrenched gender inequalities. So, I will say that using the sustainability argument is helpful because it gives us leverage to look at other dimensions, it is not just about the number of men or women beneficiaries, but it is about ensuring long-term benefits and changing mind-sets. It's really looking at what it means in terms of the people who are affected, who are we targeting. In that sense, it can help our projects to be better in terms of gender. It is not enough to create employment for women if it doesn't come with a whole set of other things.

Given the FAO's mandate and its position as a technical organisation within the UN, how much is possible to achieve and what kinds of structural issues need to be addressed for widening the approach to gender equality beyond the business case?

Clara: The problem is that if we look at the kinds of projects that FAO does, it's about creating livelihood opportunities and access to value chains and markets, or it could be about improving access to land and other resources. But then, it hardly ever goes into issues like care work or sexual and reproductive rights. And it's hard to expect that an agency like FAO would do that, although we are improving now on the issue of care work and work burden in general. But in terms of the actual project, it's not that easy to embrace a more holistic approach to women's empowerment for different reasons. Sometimes it could be funding constraints, budget lines, the kinds of projects. But it could also be simply because the counterpart is only the ministry of agriculture. So in that sense, I think it's good to use opportunities [such as the discursive leverage provided by the SDGs] to really highlight that it's not just about providing more employment opportunities or access to services, there needs to be consideration of women's work burden and the rights to their physical integrity, for instance. And I think that at the country level, these things are not always seen in a very holistic way. But not only at the country level, even at the level of FAO, so it's good to use these opportunities to remind ourselves of that. And that, for instance, FAO doesn't talk enough about women's rights in general because everything that we do should also be grounded on a human rights-based approach. It's good to put the rights agenda in there, and to use all the opportunities we have to highlight that.

Within the Asia Pacific region, Clara had been able to widen the FAO agenda by calling upon the SDGs, and also by making reference to the UN's Commission on the Status of Women (CSW), which in 2018 focused on the empowerment of rural women and girls.

Clara: That also provided an opportunity, for instance, we organised the Asia Pacific high-level meeting in preparation of CSW. So, we brought together ministries of agriculture, rural development, and gender machineries. Because the struggle is still also very much at the level of the countries similarly to what we see in our own organisations. That is, different ministries have different agendas, they don't always speak to each other, they're often in competition with each other also. And gender machineries are not that strong, to be honest. So that was another opportunity that was useful in terms of bringing different people together. And, again, in those cases, because of the SDGs and because of the background of CSW, we could actually bring together different perspectives on empowerment which is not just about economic empowerment, but it's more broadly about advancing women and girl's rights in different areas. So that was good.

Clara points out that organisations like FAO are not well equipped for this kind of thinking and approach if they work alone.

Clara: This is why it's really important when we develop projects to have on board experts who can guide in this sense, and can help project implementors and different stakeholders to look beyond what is immediately evident, and to do a proper gender analysis and stakeholder analysis, to look at power dimensions, culture dimensions and all that. And this is something that we preach a lot. I'm not sure it is always taken up. Sometimes it is, but it is not systematic yet, and this is what is needed. It is more difficult for specialised or technical agencies like FAO. I see the value of trying to work together with others, including women's organisations, or to bridge the gaps where these exist. In practice, it's not always so simple – there's also fierce competition between agencies over very limited resources.

Where she has been able to make a difference is when she moves from providing feedback on projects that have already been designed and is instead involved from the start.

Clara: I'm not always designing the project myself, I'm providing feedback on something that has already been designed, when it comes to me it is at the final stage so there is only so much you can do. The other thing is that when I'm also trying to design something myself, again I have to balance my 'ask' between what I think should be done, and what I then can do. The end result is always a compromise. But having said that, I'm not saying that I've given up or that I take a hands-off kind of attitude on whatever work or projects are ongoing in FAO. I do try and I think one of the suggestions is always to have first of all a gender specialist to do a proper analysis of whatever projects we are designing, and to have a budget for that, because otherwise we know it will not happen. And then when I do have the opportunity to be engaged from the start then I try as much as possible to use whatever additional insights I might have coming from different sources and bring them to bear on the project.

As a practical example, we are formulating a big project in Papua New Guinea with the European Union. I went there for a pre-assessment of the situation. We went to several communities in the province where the project is supposed to happen and did interviews with women's groups and civil society organisations and the big problem there is gender-based violence (GBV) and very unequal distribution of roles, so women doing most of the heavy lifting and not getting any benefits out of that. It is quite obvious that the projects will not be able to do anything directly to address GBV but the level of awareness on the problem in the country, there are some women's groups that have been active in the rural areas. One of the suggestions is to collaborate with this kind of grassroots women's organisation and also to put in the project one item that looks at women's safe access to markets and women's time poverty and work burden. So, these are obviously not going to be fully transformative kinds of interventions because the overarching goal of the project is to increase the productivity of smallholder farmers through better engagement in selected cash crops. But at least we can do some good gender analysis on how women are engaging in production and marketing, what are the constraints they face, and what could be the space for them to organise themselves in groups. Sound gender analysis has the potential to make our project so much better. The project also plans to engage other UN agencies to each contribute from their angle to the overarching project objective as well as to gender equality. So in the constraints that we have there are still opportunities for us to at least ensure the project does not harm, and that it actually benefits and empowers women. The problem is that we (gender specialists) are not always engaged from the start. This is one good example, but against this there could be dozens or hundreds that I'm not involved in. That's one of the constraints that we have.

The example Clara gives is indicative of the ways projects framed and envisioned through approaches highlight the technical management of gender equality through their integration into capitalist markets (and all which accompanies this). It shows how taking a holistic approach to knowledge, working collaboratively with relevant partners and centering social sustainability helps her identify opportunities for appropriate, if not always transformative change. The example also shows how projects are sites of ethical negotiation (Bergeron, 2016). Such negotiations 'on the ground' potentially exploit fissures in the business case for gender (improving efficiencies of women smallholders) and begin to edge towards the feminist terrain of care and solidarity (e.g. linking women farmers with GBV civil society organisations in Papua New Guinea), thus repairing the dissonance between gender as professional enterprise and feminist political ecology in practice.

Soft skills and compromise in making changes on the ground

The final theme that emerged in our discussions with Clara centred on the soft skills that must be used to bring partners and colleagues on board with gender equality, in order to deepen their engagement and avoid the limitations of box-ticking that more quantitative measures of gender mainstreaming have been accused of. The context in which Clara works throws up particular challenges are: working in a technical organisation such as FAO and its government-level counterparts where the principal disciplines of her colleagues relate to agricultural and environmental sciences, and working in a region where conservative and often authoritarian governments give limited latitude for transformative gender work.

Central to Clara's work is the building of relationships with colleagues and counterparts that she identifies as sympathetic and engaged with gender equality work, a strategy which has been identified as important within the literature on gender experts more broadly (Bustelo et al., 2016; Eyben & Turquet, 2013).

Clara: It is about finding champions, or identifying people who have really bought into the cause and working through them. So every time that I find a colleague who's more open or who's already supportive, I try as much as possible to work with them and to support what they are doing. It really has to do with talking to people.

At country level, whilst it is possible to find champions within civil society organisations and sometimes in government ministries, this kind of work requires considerable agility, and highlights a contrast between her academic research as a feminist political ecologist with a critical perspective on gender and land rights in Southeast Asia, and the care with which she must tread in her work with FAO. The subject matter of Clara's work on the gender equality implications of large-scale land deals (Park et al., 2015; Park and Maffii, 2017; Park, 2019) is well within the remit of FAO. However, at country level, the FAO's international discourse and norm-building around gender equality within agriculture and food security often confronts narrow political limits in terms of what can be said or done. What role is there here for reconciling her work in feminist political ecology, with its emphasis on gender justice, with the work of the regional gender advisor?

Clara: When I think about the work we do in the FAO, I am very much influenced by my own research. Maybe I cannot express myself as I would when I'm writing a research paper or journal article, but then I think the influence [of having to work within certain bureaucratic constraints] is not necessarily bad because it gives me a reality check of what we are doing. It also prompts me to question what I'm doing in my capacity as a gender person at country level vis-à-vis the government being the key counterpart. So, in countries in the region that creates a lot of pressures as to what I feel I should do, could do, could say.

Does this mean that issues such as the gender equality implications of large-scale land investments which have been so central to activist-academic feminist political ecology in Southeast Asia are off the agenda for her work with the FAO?

Clara: That's not what I'm saying. I'm just saying that maybe while we would say if it had to be in a report or a project document it would not be bluntly criticising the government but again there are different levels and different ways in which things can be presented and pitched. For example a global FAO report where there has been research done on the gender implications of land deals, of course, we didn't call them land-grabbing but land-based investments and agriculture investments for instance. And then it's useful always to refer to policy instruments like the voluntary guidelines or the responsible agricultural investments and to talk about the – to use the human rights language. At the country level, it's a bit more difficult. I can perhaps use a country example. FAO every four years negotiates this country program framework with the government and specifically, in some countries, it's often with the Ministry of Agriculture. In 2017 we started a process in one country and because it's supposed to be a consultative process there were several workshops and back and forth with the government and they were constantly trying to remove or to block or contest anything that was related to problems with land issues or a specific reference to things that are not properly addressed in the legislation. So it's okay to say that the legal and policy framework could be improved and that FAO can provide support in that, but it's not okay to say that the current policy framework is inadequate. It's okay to say that FAO can support in advancing gender equality but it's not okay to say that there's gender discrimination. That would not get the green light! So these kinds of things do happen quite a lot at the country level.

Thus, rather than the feminist political ecology critiques of large-scale land investments being set aside entirely, Clara brings them in, using her positioning within the FAO to give her some political space to air these issues in ways that would be denied to academics or activists in authoritarian or conservative contexts. She notes the advantage provided through the adjacency of other organisations within the UN that can be drawn on to navigate political sensitivities, whilst keeping her feminist political ecology transformative objectives at least partially in view.

Clara: In broad terms and thematic focus, my research interests could fit within FAO's work, but then it is an intentional decision that I made to try to keep the two as separate as possible to give myself some freedom to express myself, to experiment and stretch the boundaries, so that I don't feel strait-jacketed by my role here. Of course, it is easier said than done! We are talking about one person and this is why I have a periodic crisis with myself and my positionality! I am also doing research under FAO that will be published as an FAO product on indigenous people's land tenure systems in Cambodia for instance, we are designing a project together with ILO [UN International Labour Organization], but the language is obviously very different – some things can still be there, but somehow because of the comments you get, the sensitivities there are with the government, we have to tame so much, so at the end, it is half as powerful as it could be.

Whilst debate around gender experts has focused on the ways feminists can effect change within their organisations (as 'trojan horses' or 'warriors within') (Eyben & Turquet 2013; Sandler, 2015; Kunz & Prügl, 2019), Clara's 'double hat' shows how she is able to weave some of the objectives of transformative change (in this instance, addressing women's disempowerment and gender inequalities around access to resources) into her outward-facing work as well, navigating this by operating with agility, and by working strategically with other organisations (in this case, the ILO) that enable her to keep gender transformation and rights-based agendas in the frame.

Conclusion

By sharing her thoughts on her experiences, Clara poses difficult questions regarding conventional notions about the femocrat as a de-politicised subject by the bureaucracy and the establishment. Her reflections are illustrative of the reflexivity she deploys, which is shaped through her attachment to feminist political ecology, her experiences in the field, and her work in solidarity with activist-academics. In this, Clara is attentive to the exercises of power that she herself wields as a gender professional, and the agency provided by her professional affiliation with the FAO. At the same time, she is conscious that lurking wider powers define her latitudes for manoeuvre, compelling her to adopt useful institutional resources such as the SDGs to steer the boat onto a more transformative course. The discomfort and frustrations sometimes felt when navigating between feminist political ecology theory and her work as a gender professional are resolved in a desire not to give up, to use the tools her position gives her to make a difference. 'The question is…not how feminist scholars and activists can avoid co-optation by powerful institutions, but whether we can afford not to engage with such institutions' (True in Eyben & Turquet, 2013: 3). Former UNIFEM regional director Nyaradzai Gumbonzvanda once remarked: 'You need to be clear on the transformation that you seek to bring into this space. It's more than a job. It's a calling' (Sandler, 2015: 12, 13). When gender experts employ skilful ways to create transformative spaces through the politics of negotiation that make up gender mainstreaming and equality practice, they become the 'warriors within'.

References

Arora-Jonsson, S. (2014). Forty years of gender research and environmental policy: Where do we stand? *Women's Studies International Forum*, 47, 295–308.

Arora-Jonsson, S., & Sijapati Basnet B. (2017). Disciplining gender in environmental organizations: The texts and practices of gender mainstreaming. *Gender, Work & Organization*, *25*(3), 309–325. https://doi.org/10.1111/gwao.12195

Bergeron, S. (2016). Transgressing gender and development: Rethinking economy beyond 'Smart Economics' In Dinerstein, A. C. (Ed.) *Social Sciences for an Other Politics: Women Theorizing without Parachutes* (pp. 65–77). London: Palgrave Macmillan.

Boserup, Ester (1970). *Woman's Role in Economic Development*. London: George Allen & Unwin.

Bustelo, M., Ferguson, L., & Forest, M. (Eds.) (2016). *The Politics of Feminist Knowledge Transfer: Gender Training and Gender Expertise*. Palgrave Macmillan.

Çağlar, G., Prügl, E., & Zwingel, S. (2013). Introducing feminist strategies in international governance. In G. Çağlar, E. Prügl, & S. Zwingel (Eds.), *Feminist Strategies in International Governance* (pp. 1–18). Abingdon: Routledge.

Calkin, S. (2015). Feminism, interrupted? Gender and development in the era of 'Smart Economics'. *Progress in Development Studies*, *15*(4), 295–307.

Chant, S. (2012). The disappearing of 'Smart Economics'? The World Development Report 2012 on gender equality: Some concerns about the preparatory process and the prospects for paradigm change. *Global Social Policy*, *12*(2), 198–218.

Chant, S., & Sweetman, C. (2012). Fixing women or fixing the world? 'Smart Economics', efficiency approaches, and gender equality in development. *Gender & Development*, *20*(3), 517–529. https://doi.org/10.1080/13552074.2012.731812

Chappell, L. (2002). The 'femocrat' strategy: expanding the repertoire of feminist activists. *Parliamentary Affairs*, *55*(1), 85–98. https://doi.org/10.1093/parlij/55.1.85

Collins, A. (2018). Saying all the right things? Gendered discourse in climate-smart agriculture. *The Journal of Peasant Studies*, *45*(1), 175–191. https://doi.org/10.1080/03066150.2017.1377187

Cornwall, A., Harrison, E., & Whitehead, A. (2007). Introduction. Feminisms in development: Contradictions, contestations and challenges. In A. Cornwall, E. Harrison, & A. Whitehead (Eds.), *Feminisms in Development: Contradictions, Contestations and Challenges* (pp. 1–20). London: Zed Books.

Cullen, P., & Murphy, M. P. (2018). Leading the debate for the business case for gender equality: Perilous for whom? *Gender, Work & Organization, 25*(2), 110–126. https://doi.org/10.1111/gwao.12199

Davids, T., & van Eerdewijk, A. (2016). The smothering of feminist knowledge: Gender mainstreaming articulated through neoliberal governmentalities. In *The Politics of Feminist Knowledge Transfer Gender Training and Gender Expertise* (pp. 80–96). London: Palgrave Macmillan.Eyben, R. (2007). Battles over booklets: Gender myths in the British aid programme. In A. Cornwall, E. Harrison, and A. Whitehead (Eds.), *Feminisms in Development: Contradictions, Contestations and Challenges* (pp. 65–78). London: Zed Books.

Eyben, R. (2010). Subversively accommodating: Feminist bureaucrats and gender mainstreaming. *IDS Bulletin*, 41(2). https://doi.org/10.1111/j.1759-5436.2010.00123.x

Eyben, R. (2015). Promoting gender equality in the changing global landscape on international development cooperation. In A. Coles, L. C. Gray, & J. H. Momsen (Eds.), *The Routledge Handbook of Gender and Development* (pp. 515–526). Routledge.

Eyben, R., & Turquet, L. (Eds.) (2013). *Feminists in Development Organizations: Change from the Margins*. Practical Action Publishing.

FAO. (2011). *Women in Agriculture: Closing the Gender Gap for Development. The State of Food and Agriculture 2010–11*. Rome: Food and Agriculture Organization of the United Nations. http://www.fao.org/publications/sofa/2010-11/en/

FAO (2013). *FAO Policy on Gender Equality: Attaining Food Security Goals in Agriculture and Rural Development*. Rome: Food and Agriculture Organisation of the United Nations. www.fao.org/docrep/017/i3205e/i3205e.pdf

FAO (2018). *Empowering Women, Powering Agriculture. The FAO's Work on Gender*. Rome: Food and Agriculture Organisation of the United Nations. http://www.fao.org/documents/card/en/c/CA2678EN

Ferguson, L. (2019). *Gender Training: A Transformative Tool for Gender Equality*. Springer Berlin Heidelberg.

Forsyth, T. (2003). *Critical political ecology: The politics of environmental science*. Routledge.

Fraser, N. (2009). Feminism, capitalism and the cunning of history. *New Left Review*, 56, 97–117.

Harcourt, W., & Nelson, I. L. (2015). *Practicing Feminist Political Ecologies: Moving Beyond the 'Green Economy'*. Zed Books.

Kunz, R., & Prügl, E. (2019). Introduction: Gender experts and gender expertise. *European Journal of Politics and Gender, 2*(1), 3–21. https://doi.org/10.1332/251510819X15471289106077

Miller, C., & Razavi, S. (1998). *Missionaries and Mandarins : Feminist Engagement with Development Institutions*. London: Intermediate Technology Publications in association with the United Nations Research Institute for Social Development.

Park, C. M. Y. (2019). 'Our lands are our lives': Gendered experiences of resistance to land grabbing in rural Cambodia. *Feminist Economics, 25*(4), 21–44.

Park, C. M. Y., & Maffii, M. (2017). 'We are not afraid to die': Gender dynamics of agrarian change in Ratanakiri province, Cambodia. *The Journal of Peasant Studies, 44*(6), 1235–1254.

Park, C. M. Y., White, B., & Julia, J. (2015). We are not all the same: Taking gender seriously in food sovereignty discourse. *Third World Quarterly*, *36*(3), 584–599.

Parpart, J. L. (2014). Exploring the transformative potential of gender mainstreaming in international development institutions. *Journal of International Development*, *26*(3), 382–395. https://doi.org/10.1002/jid.2948

Prügl, E. (2015). Neoliberalising feminism. *New Political Economy*, *20*(4), 614–631. https://doi.org/10.1080/13563467.2014.951614

Resurrección, B. P. (2017). From 'women, environment, and development' to feminist political ecology. In S. MacGregor (Ed.), *Routledge Handbook of Gender and Environment* (pp. 71–85). Oxon: Routledge.

Sandler, J. (2015). The 'warriors within': How feminists change bureaucracies and bureaucracies change feminists. In R. Baksh & W. Harcourt (Eds.), *The Oxford Handbook of Transnational Feminist Movements* (pp. 188–214). Oxon: Routledge.

Sandler, J., & Rao, A. (2012). *Strategies of Feminist Bureaucrats: United Nations Experiences* (Working Paper No. 397). Sussex, UK: Institute of Development Studies.

Skard, T. (2009). Gender in the malestream—Acceptance of women and gender equality in different United Nations Organisations. *Forum for Development Studies*, *36*(1), 155–197. https://doi.org/10.1080/08039410.2009.9666429

UN Women (2018). *Turning Promises into Action: Gender Equality in the 2030 Agenda for Sustainable Development*. New York: United Nations. https://www.unwomen.org/en/digital-library/publications/2018/2/gender-equality-in-the-2030-agenda-for-sustainable-development-2018#view

World Bank (2006). Gender action plan: Gender equality as smart economics. http://documents.worldbank.org/curated/en/295371468315572899/Gender-equality-as-smart-economics-a-World-Bank-Group-gender-action-plan-fiscal-years-2007-10

World Bank (2011). *World Development Report 2012: Gender Equality and Development*. Washington DC: World Bank. Retrieved from: http://documents.worldbank.org/curated/en/492221468136792185/Main-report

5 Challenges and dilemmas of integrating gender in the field of environment and development at SEI

Metrics and metaphors

By Andreea R. Torre
In conversation with Natalia Biskupska, Maria Escobar, Laura Forni, Emily Ghosh, Ha Nguyen, and Lisa Segnestam

The growing relevance associated with gender equality in international policy-making, including in the Sustainable Development Goals, the Sendai Framework for Disaster Risk Reduction, and the Paris Agreement on Climate Change, has raised new demands for gender expertise in areas where engagements with gender experts had been few. As a result, and following donors' interest in gender mainstreaming, the involvement of gender researchers is becoming more common in projects and programs ranging from water and sanitation, climate change, disaster risk reduction among many others.

Yet, although gender has been recognised within environmental research and policy, and conceptualisations about gender and environment have also dynamically evolved since the late 1970s and 1980s (Arora-Jonsson, 2017; MacGregor, 2017), the actual integration of gender perspectives and analysis within research and sustainability projects is still a challenge. Lack of awareness or technical 'know-how' are among the reasons for this inertia. As some of us in this chapter mention, more conventional environmental research had often directed its focus towards environmental sustainability, for instance, or inequalities on a larger scale (e.g. urban spaces with fewer resources) neglecting the relevance of understanding smaller scale experiences and relationships.

The increasingly common focus on addressing climate change impacts in both research and programming for development, adds further layers to the positioning of gender experts within the broader spectrum of social researchers and technical specialists called upon to help ensure that the ecological and social unsustainability of current development paths is addressed and that equitable strategies are proposed to redress this trajectory. In this case gender professionals have to engage with a common (neoliberal) framing of environmental issues as primarily an aspect of natural science, or techno-scientific problems often calling for technical solutions (e.g. engineering of shock-resistant crops), market-based mechanisms and regulatory interventions (e.g. emission trading) and the intrusion of non-state actors (e.g. corporations) to mitigate or to adapt to environmental hazards and climate change impacts.

This chapter develops through a conversation with fellow gender specialists and researchers who have placed gender at the centre of their environment-related studies and policy engagement in the different centres of the Stockholm Environment Institute (SEI) across Asia, Europe and the USA. Marisa Escobar, Andreea R. Torre, Laura Forni, Lisa Segnestam, Ha Nguyen and Emily Ghosh reflect on the challenges of assiduously working for bringing gender in, and keeping it central, to the realms of environmental science, capacity-building activities and policy engagement.

Those challenges are multiple and take place at different scales. They relate to complexities of situated geo-political identities of 'being a woman scientist' and to differing disciplinary and professional backgrounds as well as to the nature of work that involves navigating encounters between epistemologically masculine fields which often overlook the micro-scale of human and more specifically women's experiences. The reflections in this chapter are hence rooted in the never-ending dilemmas about self-identifying and being identified as gender professionals, the ontological and epistemological tensions between natural and social science researchers, as well as the politics and practice of gender work in a historically technical, and increasingly managerialist field.

Self-identifying and being identified as gender experts

Reflexivity and identity positioning are processes that allow researchers to connect to their work experiences and grasp the intricate relationships, intrinsic tensions, and winding paths of learning that often shape such experiences. Calls for thorough analysis of the working experiences of gender experts in different research fields and institutional contexts are not new, and responses have developed especially in the field of international development (Prügl, 2013; Ferguson, 2015; Bustelo, Ferguson & Forest, 2016). In keeping with this ongoing work, a self-reflexive exercise focusing on our experiences and shifting, troubled identities as gender professionals in the specific field of environment and development kicked off our conversation while connecting through cyberspace in the Skype chatroom.

We reflected on the 'roads' to our current positions within SEI and on what it takes to be identified as gender specialists. Kunz and Prügl (2019, p. 3) highlight how gender experts, while defining a new body of knowledge and marketing their ideas in multiple spaces, have also established a new profession. They 'implement gender mainstreaming, offer specialised training, spread gender-equality considerations throughout organisational structures and seek to bring about social change through the implementation of gender-sensitive projects'. What has strongly emerged from our conversations is the diversity of experience, of understandings about being a woman scientist and gender specialist.

In our roles as SEI researchers, some of us feel caught between the conflicting positions of the 'gender specialist' expected to provide technical advisory services and the feminist researcher, who is concerned with transforming the workings of unequal power. Others reflected on their evolving position from technical scientist to 'gender champions' in their technical fields of water and energy, or on their early steps towards a future, aspired profession. Finally, we observed how power dynamics and intersecting social variables of place, class, ethnicity, race infuse a political twist to our work on gender and environment in development contexts. The variety of SEI office locations, our areas of expertise, academic and professional paths, allowed for insightful stories to be told.

Andreea: How would you define gender professionals and do you consider yourself one of them?

Lisa: I have a background in economic history with a focus on intersectionality and vulnerability to climate change. Now I can consider myself a gender professional since I've spent my last 10 to15 years working on these issues although I'm not a gender professional by training. It's more about putting together different pieces of theory and practice.

Marisa: I don't consider myself as a gender professional, but I think that through my involvement with the Gender and Social Equity program [GSE] at SEI, I've been a bit of a champion on gender issues wherever I go. I have been involved in incorporating gender in modelling work through the GSE program and in the 'Water Evaluation and Planning' [WEAP] tools. In workshops where I go, even if many are conscious about these issues, I think someone needs to talk about them. Someone needs to bring those issues to the table. To say, for example, hey, this is not a balanced workshop, because it is not 50/50. So, I think I bring that voice with more arguments to make sure that people are aware.

Andreea: Few years ago, I wouldn't have considered myself a gender specialist; most probably because of my background which is not strictly in gender studies. I got closer to the gender and development field though the work that I have done. The gender specialist identity has emerged and was shaped through my first academic position and through engagement with feminist organisations and development agencies working on women's issues in the South Pacific. Yet at that time I would identify myself and I was identified by others as a sociologist and lecturer and researcher in development. Now, by being part of the Gender, Environment and Development cluster at SEI, this identity [gender specialist] has become more solidified. Personally, I'm interested in understanding that messiness of the world around us, and where gender comes into it rather than just focussing on gender as such. I'm interested in all those dynamics and how they come into place. So maybe, technically I may be defined as a gender professional, but I feel that I am more than that.

Ha: I have a background in gender and development studies and I previously worked as a gender focal point providing support to various projects on agriculture and rural livelihoods. For me gender professionals are those who make a living by offering services and advice on gender. So, in that sense I consider myself as a gender professional. I did not consider myself a gender professional in my previous work – although I was trained in gender – just because of the nature of the projects I was working on and which did not have much of a focus on gender. I do consider myself a gender professional now as I work on projects on gender and I do consultancies on gender. This has become my focus.

Natalia: I would have described the gender specialist as a standalone person who works on gender. Whatever that means I wouldn't know – a person working on those issues. But now that I'm involved in this area, I've realised that gender professionals are people working in certain sectors on the same issues as others but explaining things in a unique way and looking at gender aspects of those issues. So rather than being a standalone person, I see them as integrated in those sectors. The effectiveness of that work is the challenge; trying to bring down barriers and being heard in those sectors. I think I'm a want-to-be gender professional but still have a lot to learn.

Emily: Through my involvement in the Gender and Social Equality program, I am currently working on research related to gender issues in the energy sector. I would not consider myself a gender professional at this time as this area of research is still new to me. I see a gender professional in our field as someone with knowledge of the history and issues related to gender within various aspects of sustainability and development and with a focus on gender in a large portion of their research. I think of myself more as a gender champion within the workplace, pushing for more awareness on gender issues and gender inclusion in the projects we do.

Laura: I don't consider myself a gender expert in the wider scheme of things. I'm a technical scientist with a background in economics and have worked on water resource management. Through the GSE program I started exploring gender in the modelling work considering the mobility aspects in relation to gathering water and other projects where funders have asked for gender components to be included. Hence, I do consider that I became a water-based gender professional or a professional at the intersection of gender and water resource management. This is because of all the reading and work that I have done in the last two years. Moreover, as the chair of the gender and diversity committee in the SEI US, I helped developing policies for the institute regarding equity, diversity, gender and inclusion. And I think that because of the engagement within this process I became a lot more aware of gender in the workforce. As we explore the inequalities in our scientific work, it is inevitable to become aware of the inequalities outside the science. Such is for example the role of, or the lack of, women in decision making and scientific panels. As well as of the biases that are presented in the societies we are located in or we work with internationally.

Rather than voicing merely personal narratives, the relevance of this discussion is to be found in the ways it allowed us to approach connections: between the personal, the social, and the political as well as the linking with the 'historic situation', 'social structures', and 'moment of experience' (Denzin, 1997: 39) which shape career paths and professional identifications. 'There is a clear argument for intersectionality emerging within the group', as Lisa observes anticipating matters that will be picked upon later in this chapter. The emerging narratives tell a story of a process of 'getting into the work' of the gender professional. It is about gradually growing into an identity as a researcher, a scholar, a professional, and a champion of gender equality in the environmental field.

It also tells us that what makes a gender expert is an assemblage of instruments (theory, background, mainstreaming skills etc.) making up, at various degrees, a 'toolkit' of expertise. This is a dynamic process. This is a process of becoming those different identities and concomitantly of overcoming the often-encountered *epistemic injustices* (Fricker, 2007) of being women scientists and/or of engaging and bringing in the gender analysis in technical fields.

Conversations below unpack on one hand the positioning of women technical scientists and on the other, that of gender specialists in the natural resource management fields and environmental organisations. While Laura, Emily and Marisa stress the trivial visibility that women scientists are given because of their gender identity, Ha's reflection pushes the discussion farther to uncover the lack of legitimacy gender studies/analysis suffer as a science and as a discipline against the epistemic authority of technical sciences. Yet in both cases, as Natalia pointed out earlier, rather than shaping a stand-alone profession, we work towards integrating our work and understanding in various, diverse sectors of environment and natural resource management. And in both cases, there are barriers to overtake and knowledge to establish.

***Technical women in technical fields*: the everlasting struggle for epistemic credibility**

In her analysis of forms of epistemic injustice, Miranda Fricker (2007) defines the experience of being devalued as a knower as 'testimonial injustice'. She notes how individuals can face a lack of epistemic credibility because of a collectively shared identity preconception. In these circumstances a woman can be denied recognition. She can be marginalised as a knower because of a systemic prejudice regarding her gender. Such is the condition of being a woman scientist in a technical field constructed as 'masculine' through consolidated norms and practices of knowledge production.

The rise of modern science in Western Europe has been associated with dynamics of professionalisation of science and its exclusion from the personal/domestic realm – a *locus* for the production of science in earlier times (Abir-Am & Outram, 1987). This shift encouraged the collective identification of men with the external, the public, the rational world while ascribing women to the emotional dimension of their nurturing roles and 'soft' skills. A gender binary was established to support assumptions about a 'structured incompatibility of women and science – and between women and scientific rationality' (Seager, 2017: 38). Commenting on the state of science, as well as environmental politics, in the late twentieth century, Merchant (1980, 2006) points to the persistent belief in 'objectivity' and decontextualised knowledge among Western scientific communities. A belief that is responsible for the limited questioning of grand narratives underpinning the *testimonial injustice* which still prevents women scientists from fully engaging in the epistemic practices of science.

Laura, Emily and Marisa dialogue here around the troubles and injustice women scientists must overcome in technical settings. They recall accounts of their own experiences and of other women colleagues not being fully accepted as scientists in a field which is epistemologically masculine. They observe how women are either unwelcome or, when in, they are 'invisible' knowers and expected to perform stereotypically 'domestic' and 'feminine' tasks. When that assumed 'performativity' is not exercised by women professionals, 'accommodating gender in a project turns somewhat uncomfortable' – as Marisa's words brilliantly point out at the end of this exchange.

Marisa: In many instances of my professional work I have faced the all too common reality of being the only woman in the classroom or in project meetings. As such, I had to overcome some of the cultural biases that prevent women to occupy these roles. Such biases come in many subtle forms and on a daily basis. I encounter mansplaining, or associated forms of it, quite often. These events of either interrupting my explanations, giving me unsolicited advice or ignoring my opinion, disturb the regular flow of a workday, keeping me down and requiring time to recover. This happens at meetings where my voice or ideas seem to not be heard.

In other occasions, what happens is that my emails are not read or are ignored, requiring me to go through a male colleague to produce the e-mail I need the counterparts to read. By doing this I get my work done, but of course it requires more work and frustration. Difficulties may arise when participating in management meetings. In those cases, there is a level of assumed knowledge and a level of comfort from, mostly male, colleagues in higher ranks that is counterproductive for an effective transition into increasing levels of seniority.

Laura: Not long ago, I attended a workshop where capable female professionals, including myself, participated. One of the male presenters made a joke about bringing water pumps to low income communities and the potential of making women fatter because they wouldn't have to walk to collect water anymore. In spite of the distasteful comment, several laughed. Only a few perceived the joke as offensive and only admitted it later when I brought it up. My impression was that, overall, it was acceptable. It was 'just a joke'. In a culture that is so immersed in the 'man's way of thinking' it is hard to create a safe space for women to feel valued.

Emily: Yes, unfortunately, more often than not, a woman will be asked to do the admin tasks in a project 'because they are good at it' and because this at least gives them an opportunity to work on a project in some form. However, this affects future opportunities to engage in technical work. There needs to be greater emphasis on giving women a seat at the table in decision-making processes. By doing so, not only a diverse set of perspectives is included, but this can help recognise and avoid having gender-unequal roles in a project.

However, greater gender inclusion in a project does not always guarantee gender inclusive outcomes. For example, society deeply undervalues the contributions of household work, such as cooking and cleaning, in an economy; work that is predominantly done by women in every country in the world. As a result, energy policy decisions may be skewed to prioritise industrial energy demands (i.e. male-dominated fields) rather than ensuring, for instance, safe and reliable electricity supply to households. This is why advocating for gender analysis in projects, alongside the typical environmental, social and economic analyses, is paramount.

Marisa: Not all projects require gender integration and when this is not explicit in the projects, there is an additional challenge in order to motivate gender considerations with the funders and project team. This challenge is twofold: first, why add a gender layer to a water analysis project that is already complicated in itself; and second, the women team-members often play, or are expected to, an administrative support role.

This second point is important because when gender considerations are brought to the surface as part of the project, it generates an uncomfortable situation where the role of the woman needs to be shifted from administrative support to technical expert. This situation potentially leaves the men in the team without the appropriate administrative support and perhaps with the burden of sharing some of that work. At this point, accommodating gender in a project turns somewhat uncomfortable and with the potential to harm the politics of managing the project.

Gender analysis: sitting on the margins

Having walked a different path, that of the researcher and gender specialist in an environmental organisation like SEI, Ha adds her views to the discussion above. They highlight the knowledge politics encouraging forms of epistemic authority of the technical sciences and engendering spaces where gender analysis lacks recognition and legitimacy. In those spaces, where paradigms of scientific objectivity are dominant, problems are seen as mainly technical and hence in need of technical solutions that natural sciences only appear to be able to provide.

Furthermore, the widely spread essentialist association of gender studies with 'women issues' has further contributed to relegating gender analysis to the epistemic margins of environmental research and policy. Gender is deemed peripheral, if not irrelevant, to the technical efforts to correct environmental damage and efficiently manage natural resources, as Ha narrates below. Gender specialists are expected to show up armed with *tools* (e.g. techniques and measurement instruments) to 'fix' rather than transform unequal and unjust power relations which should be at the heart of the gender agenda (Cornwall, 2007).

Ha: It's a 'half-way commitment'. In my previous position as gender adviser, efforts were made to set up an organisational structure that aimed to support gender mainstreaming. This structure included assigning a gender lead, setting up a gender working group, conducting gender scoping in the organisation, developing gender mainstreaming policy and minimum standards. These were often demonstrated as the organisational commitment to gender equality, but they were lifeless. No further efforts were made to ensure that the policy and guidelines were followed. No one was held accountable for not following the policy, responding to gender advices, nor delivering expected results regarding gender.

'It's something to show off': It was not challenging to request budget for gender analysis, disaggregated beneficiaries or capacity building for women as long as those activities did not affect what the programs were set out to do. After all, having gender budget was something to show off. I encountered huge resistance (e.g. excuses, disengagement…) once gender power relations were introduced and commitment to gender equality implied or required changes in program design, approach or implementation process. I think many program folks preferred to conceptualise gender as sexual dichotomy where they could disassociate themselves from the development context, rather than power asymmetry where they might well be part of the (gender inequality) problem.

'I am a dog without teeth': This is a phrase that one of my gender specialist colleagues called herself. I felt exactly the same. As the gender lead, I was requested to review program proposals and strategies, but nobody was held accountable for responding to our recommendations to strengthen the gender dimensions. It was no more than ticking the box.

Overcoming the 'troubles' (1): the gender expert

Ha tells us about her work with Geographic Information Systems (GIS) experts and explains how the troubles related with integrating gender are caused by the lack of awareness of the importance of gender issues in their project, and how it is marginalised as a field of expertise. There would be more receptivity if there was consensus around social justice as the main purpose of this exercise. One way to integrate gender, Ha highlights, is for ourselves – technical and social specialists, planners – to critically engage and reflect on our own positioning (as development professionals, researchers, technical scientists etc.) within the very game of power relations.

Ha: I was tasked to support a climate forecasting service developed by GIS experts and technicians to reach out and benefit women and men farmers in a southern province of Vietnam. We had conducted a gender assessment to point out problems faced by some groups of farmers, particularly ethnic minority women and their communities, in accessing climate services and using climate forecasts in livelihood planning. The heart of the problem was the centralised power of the state that determines access to and use of different types of climate information, directs agriculture planning and regulated water resource allocation aligned with economic growth objectives. As a result, they did not target small-holder ethnic minorities, women and their livelihoods because these did not fall under those objectives. These groups could not use these services to adapt their livelihoods to climate change.

What else does this experience also tell us? One: there was an assumption that once you hire a gender specialist 'to do gender', gender issues will be addressed but not at the expense of altering the entire 'technical exercise'. Two: in technical areas of expertise such as GIS, gender is not considered a science. Resources are mostly committed to technical interventions rather than addressing power-related inequalities and social justice.

Andreea: So, what I'm hearing here is that doing work on gender is an instrument for bringing in change by transforming power relations. Am I right?

Ha: We need to change our perception and ways of addressing gender and social inequality. We have to recognise that we are part of the problem and also the solution. For example, back to the case that I mentioned earlier, if the GIS program team does not change the way they design and deliver the forecasting tool and service, they will serve the existing structure that produces unequal access to climate information and undermines coping strategies of the traditionally marginalised grassroots groups. Being aware of our role in producing inequality might create the urge to engage with and address it.

Overcoming the 'troubles' (2): the technical scientists

While Ha's proposal of overcoming troubles connects firmly with her background and current positioning as gender specialist, other proposals take different forms. Emily and Laura provide a detailed account as technical scientists engaged in integrating gender in projects. They see the relevance of sharing knowledge on gender with their peers at the outset of project work, as well as of engendering a change in methodology which focuses on the individual and her place at the centre of contextual dynamics and relations of power.

Andreea: How are our experiences of integrating gender into our work on environment different based on your areas of work at SEI?

Emily: In the energy sector, we are generally finding that more project terms of references' specifically request that a gender analysis be conducted as part of the project. We often review the health, social and economic impacts of the various climate mitigation measures that are analysed in a project. But additionally, we are now asked, what are the gender implications of that? Hopefully in the future, GSE concerns will no longer be an afterthought, but an integral part of energy and climate mitigation policies.

Andreea: Can you say a little bit more on that? How do you actually do it?

Emily: Well, there's been a lot of discussion in the last year on how to analyse the gender impacts of different climate and energy policies. Energy planners are often unaware of the gender implications of their decisions. So, what I have been working on recently is to put together a detailed overview of gender issues in the different aspects of energy planning and create a training module for LEAP for energy planners to learn about gender issues through training exercises. Depending on the project, this module is something that could be referenced when doing a study of various policy options.

Major gender and energy issues in rural areas of developing countries revolve around the type of energy consumed within a household, and how that affects women who predominately do household work. When we are looking at household energy demand, there's a list of concerns related to gender that need to be considered related to time burdens, pollutant exposure, and various health and safety concerns. On the energy supply side, there are concerns related to displacement, for example from hydropower dams, or pollution from coal plants, and how that can potentially disproportionately affect women.

We looked at those issues in one project in Jamaica which specifically asked for a gender analysis. In this project, we identified a number of climate mitigation options for the country and then analysed the gender impacts from a health perspective, a safety perspective, as well as on general productivity and lifestyle. While this was more of a qualitative analysis, we are now also developing methods to quantify the impacts so that our energy planners can say: 'Okay, this is the time lost for a person that is spending 10 hours a week collecting fuelwood instead of using LPG or electricity for cooking – time that could be used for income-generating activities or school.' We would like to help energy planners quantify those impacts and put numbers to the decisions that are made. Yet, sometimes the data available for quantifying gender impacts is limited.

Andreea: Do you feel that there is greater demand for quantitative data rather than, as you say, more qualitative understanding of the issues?

Emily: I think that there is room for both. Looking at impacts from a quantitative perspective can be useful, especially for technical energy planners who look at numbers to understand the scale of a problem. But qualitative analysis also has a place in the energy sector.

Laura: I see the value of quantitively addressing gender issues, although, in these cases, the gender analysis often gets stuck in counting people. I believe we need to go beyond that, for example in addressing the impacts of water distribution and how they differ by gender. One example is looking at the 'opportunity' cost of women's labour in collecting water when they don't have access to water in their premises. So that gives a quantitative sense of one of the impacts of unequal access to water and the social pressure put on women to collect water elsewhere when there is no access to it at home.

Also, in terms of achieving equality, I think empowerment is key. I see the value of showing gender differences in the numbers to inform policies, but also to create the awareness around gender aspects that need to be considered when we examine how water is managed within a watershed. We need more tools though to continue addressing the relevant linkages between gender differences in water management impacts, not just technical tools but also qualitative tools that can help us understand and identify the main aspects the model needs to focus on.

A change in methodology

With a specific focus on the development of modeling tools for water and energy planning, Laura refers to a change in methodology which positions individuals in the context of power relations and differentiated structural and contingent everyday challenges. She tells us how gender mainstreaming in long-term resources planning modeling tools could be achieved through a *process* that deepens our knowledge of both *metrics* and *metaphors*; a process able to provide an all-encompassing picture of the lived (and unique) realities, and hence situated and embodied experiences and knowledge (Haraway, 1988), of women and men in challenging contexts.

Laura: There is a large constituency of support for gender equality and women's rights from researchers, policy makers and funders. The current challenge is, however, how this constituency can develop the metrics and metaphors that provide evidence so practitioners can support making a difference at the local, regional or national levels. The process of gender mainstreaming towards gender equality therefore needs to be incorporated within a framework that researchers from the various fields can adopt in their analytical work.

Modeling tools for water and energy systems planning have supported resource managers and policy makers at the various governance levels in formulating decisions for a sustainable future. This process is focused on the identification of challenges that decision-makers face, which drives the metrics evaluated in the model, and the type of output analysed when formulating a decision. Assisting water managers in incorporating gender equity aspects in their future long-term planning, ensures that current inequalities are not perpetuated.

The SEI GSE program has developed a guidance document for mainstreaming gender in modelling activities and a training tutorial for an energy and water systems planning tools. However, while having a tutorial for modelling activities opens the door for researchers to evaluate gender aspects, the tutorial by itself is not enough. The process of mainstreaming gender within modelling-based research needs to incorporate the metrics, evidence, and practice from the identification of challenges to the evaluation of outcomes. That way, gender aspects are part of the key metrics to be evaluated in the model, and researchers can effectively evaluate them in the modelling tools. Mainstreaming gender needs to create a process that informs the metrics and the dialogue of policy makers to ensure that change happens. Having a modelling platform and a technical process designed to incorporate gender aspects in resource planning mechanisms facilitates a process towards a gender equal and sustainable future.

Yet while gender mainstreaming efforts at SEI are well accepted and benefit from active support, it is in projects that it is harder to include. Often funds are not available to support gender inclusion in the type of projects we have. Fortunately, lately there have been some changes. My colleagues and I do as much as we can within project objectives and budget to promote input from women in workshops and trainings. Yet it could be more impactful if the project includes a gender component in the analysis to support the exploration of what are the data, model, metrics that we need to consider to address gender differences in water access, water management, and adapting to water related challenges such as climate change impacts on water resources. It would be also important to have funders supporting aspects related to gender equality and women's empowerment in capacity building activities, workshop developments and decision-making processes.

Lisa: What do you consider as good data that will make the case for integrating gender into the environmental program or project you are jointly working on with other technical colleagues?

Laura: Demographic data or resource demand information disaggregated by gender is something that can help quantitative analysis. But we should not stop there. I think that the main challenge is the opportunity cost women face when they don't have direct access to water and they need to walk for miles. In that case, disaggregated water use data may not be as relevant as the time spent in getting water. In the model, when a water access strategy is implemented, a variable can be created to estimate that time, based on water supply estimates. This is an example of looking at the opportunity cost in having to walk miles to obtain water. Alternatively, an estimation of that time can be evaluated as the trade-off in time dedicated to educating girls and women.

But it is also about working with a different approach. The framing (ontology) around water management, and the key metrics that define satisfactory management, needs a more holistic and distributed approach that can help design new methodologies on water resources research. For example, a case study developed in Bolivia where we disaggregated the demands and were able to capture inequalities that were hidden in the standard models. Another example is a project in Cambodia where SIDA's framework for poverty analysis informed a water resources planning model to address inequalities. By using a different approach, we are working to ensure that sustainable water management happens in tandem with efforts to achieve social equality, economic growth, and poverty reduction.

Marisa added to the discussion. While highlighting the relevance of including more women technical experts, her reflection uncovered the gender politics of gender work.

Marisa: My experience integrating gender has evolved in the last few years. Initially, the development projects I worked on did not mention gender, or when they mentioned it, they did not keep track of the progress thoroughout the project to see how gender could be integrated at different stages. Another delicate component of bringing gender into projects occurs when big consulting firms manage development and environmental projects. In those instances, work can be presented as gender sensitive on the front page of the project, yet gender issues remain not addressed internally. Through the gender initiative in SEI I have been able to make time for a focused effort about how to integrate gender into our water resources modelling work. As a result, the relevance of integrating gender into the modelling work and of including women as technical experts in the analysis is established. With these two key components, and with particular strategies to include them, we can now more explicitly include gender considerations into our work.

I found it essential that there is at least one woman in each of the technical teams. The importance of having that woman is that she can embody the importance of gender considerations. This is particularly significant in specific contexts where women are persistently left out of the technical work. In those cases, the gender transformational work needs to go beyond the project itself and address societal unbalances. During the implementation of a current project in Bolivia, for example, one of the women representative of the team has proposed visiting schools. The motivation behind those visits was to make girls understand, at a very early age, that women can aspire to and undertake technical careers, such as engineering. Structural change, and power unbalances, need to be addressed at different levels and girls in schools need to see other women as role models when they are young enough to imagine a future when they can also be in that position.

Some concluding remarks

Thinking through our experiences, a picture of diversity but also of intersecting paths has emerged. This diversity related to factors ranging from disciplinary backgrounds and the cultures and geographies of work. While earlier it seemed as if a clear distinction between the professional paths and perspectives of the *technical scientist* and *the gender specialist* was shaping up, this distinction eventually blurred when SEI colleagues began to share their stories of 'want-to-be' gender professional, of 'becoming' gender experts and/or gender champions and the challenges and dilemmas in their work.

Furthermore, the search for transformative approaches to integrate gender in environmental and natural resources fields was also central to our discussion showing a common struggle despite our different ways of knowing and hence different methodologies of work. But all the more important, being reflexive about our positions – and realising inequalities in our own positionings – has allowed us to get a better sense of the broader, and at the same time deeper, common meaning of our work 'as someone oriented not only to interpret the world but to change it' (Pante, 2014: 82 after Marx, K. 1888 Thesis Eleven in 'Theses on Feuerbach').

Connections beyond our field of work *per se* were also revealed by our considerations on the use of a critical gender lens that enables us to recognise historical and contemporary oppressions, patriarchal limitations in our workplaces and societies we live in, social and personal gender biases. Both in praxis and in theory, the mainstreaming of gender frameworks and gender analysis needs to account for the 'breadth and specificity of oppressions' (Bartlett, 1989: 949) lived by different women and men. The need for intersectionality, as Lisa reminds us, not always privileged by environment and development work, was highlighted here for there is a risk that existing inequalities are ignored and/or exacerbated if multiple, intersecting markers of structural and contingent discrimination are ignored. Those factors are often not captured if tools for gender analysis do not go beyond the use of gender as a variable and lack sufficient consideration of power relations.

Finally, our conversations highlighted how an approach to climate and environmental change and its impacts which sees problems as mainly technical, tends to acknowledge and rely on traditional natural sciences only. Finding its heritage in the developments of the nineteenth century, this approach brought to light the relationship between science/scientific knowledge, economic growth and the expanding market economy. What we hence witnessed was the rise of a 'culture of knowledge' that has taken the scientific rhetoric of 'evidence' into the wider public sphere while depoliticising both knowledge and intervention. Yet accounts of 'changing framings and metrics' around water management for instance, exposed an underlying shared ontology for social justice and its potential to yet again blur disciplinary distinctions and bridge our knowledges.

References

Abir-Am, P. G., & Outram, D. (Eds.) (1987). *Uneasy Careers and Intimate Lives: Women in Science 1789–1979*. The Douglass series on women's lives and the meaning of gender. Brunswick-London: Rutgers University Press.

Arora-Jonsson, S. (2017). Gender and environmental policy. In S. MacGregor (Ed.), *Routledge Handbook of Gender and Environment* (pp. 289–303). Oxon & New York: Routledge.

Bartlett, K. T. (1989). Feminist legal methods. *Harvard Law Review*, 103, 829–888.

Bustelo, M., Ferguson, L., & Forest, M. (Eds.) (2016). *The Politics of Feminist Knowledge Transfer: Gender Training and Gender Expertise*. Basingstoke: Palgrave.

Cornwall, A. (2007). Revisiting the gender agenda. *IDS Bulletin*, *38*(2), 69–77.

Denzin, N. K. (1997). *Interpretive Ethnography: Ethnographic Practices for the 21st Century*. Thousand Oaks, CA: Sage.

Ferguson, L. (2015). 'This is our gender person': The messy business of working as a gender expert in international development. *International Feminist Journal of Politics*, *17*(3), 380–397.

Fricker, M. (2007). *Epistemic Injustice: Power and the Ethics of Knowing*. Oxford, UK: Oxford University Press.

Haraway, D. (1988). Situated knowledges: The science question in feminism and the privilege of partial perspective. *Feminist Studies*, *14*, 575–599.

Kunz, R., & Prügl, E. (2019). Introduction: Gender experts and gender expertise. *European Journal of Politics and Gender*. Special Issue: gender experts and gender expertise, *10*(10), 1–18.

MacGregor, S. (2017). Gender and environment: An introduction. In S. MacGregor (Ed.), *Routledge Handbook of Gender and Environment* (pp. 1–24). Oxon and New York: Routledge.

Merchant, C. (1980). *The Death of Nature: Women, Ecology, and the Scientific Revolution*. San Francisco, US: HarperCollins.

Merchant, C. (2006). The scientific revolution and the death of nature. *Focus – Isis*, *97*, 513–533.

Pante, B. L. P. (2014). Female researchers in a masculine space: Managing discomforts and negotiating positionalities. *Philippine Sociological Society*, 62, 65–88.

Prügl, E. (2013). Gender expertise as feminist strategy. In G. Caglar, E. Prügl & S. Zwingel (Eds.), *Feminist Strategies in International Governance* (pp. 57–73). London & New York, NY: Routledge.

Seager, J. (2017). Rachel Carson was right – then, and now. In S. MacGregor (Ed.), *Routledge Handbook of Gender and Environment* (pp. 27–42). Oxon and New York: Routledge.

Part 2

The politics of knowledge in environment and development realms

6 The politics of feminist translation in water management

By Bernadette P. Resurrección and Rebecca Elmhirst
In conversation with Seema Kulkarni and Margreet Zwarteveen

Feminist encounters with technical water professionals such as, for example, engineers, modelers and bureaucrats bring into view fundamental questions and differences in approaching and understanding the use and management of water. In the field of irrigation, technical water professionals are trained to conceptualise and model water systems with the goal of optimising water productivity for agro-industrial purposes. By contrast, feminist researchers are more concerned with the power-laden ways water is allocated, noting that water gives life and is part of the world's 'commons'. They explore how water can be accessed and distributed more fairly and decision making could be more inclusive. They therefore emphasise questions of social justice in relation to water's multiple uses and stakeholders.

In conversations with Seema Kulkarni and Margreet Zwarteveen – who are feminist researchers and activists in rural waterscapes – they share about how they have tried engaging technical water professionals and bureaucrats in transformative forms of feminism. They explain how these attempts are marked by a distinctive knowledge politics, as among others manifested in contrasting water use ontologies: those of water productivity and efficiency on one hand, and those of justice and equity on the other. These contrasts and the knowledge politics through which they are played out mean that relationships between technical water professionals and bureaucrats and feminists always require translations. In the conversation, they consider how to best engage in and navigate the politics of these translations. In their experience, the oft-used tactics of calling for 'integrating' or 'including' gender issues in water policies and projects is not the most fruitful pathway, despite the many 'success stories' disseminated by development donors based on the assumption of synergies between water resource improvements, gender equality and women's empowerment. In the experience of Seema and Margreet, integration of gender in existing projects or calls for more participation of women in existing water decision-making processes and bodies may create dangerous distortions and dilutions of feminist and social justice concerns. Instead of integration or inclusion, a strategy of cautious juxtaposition that consists of conscious strategies to learn to acknowledge differences and disagreements may be more promising.

Personal and professional identities

Our starting point in this conversation is our own positionality stemming from our own work on gender in rural waterscapes. Seema and Margreet have worked on and in irrigation in South Asia and elsewhere, with Seema having a particularly long experience in Maharashtra. Both of them have been involved in different projects to address gender inequalities or include women in irrigation projects. In talking to them, we note different places of feminist engagement with watery concerns. Where Margreet has been particularly engaged in dialogues in universities with technical colleagues and irrigation specialists; Seema has had most interactions and negotiations with local water authorities and has a lot of experience in directly engaging with landless and scheduled caste women and men. It is clear that the dominant terms in which the use and management of water is conceptualised in professional irrigation contexts are those of efficiency and productivity. A main pre-occupation of many irrigation professionals is the avoidance of waste – with waste often being defined as all water that does not enter into a market equation, or that flows into the sea without being useful to man (see Gilmartin, 2003). According to Seema and Margreet, the challenge of navigating and transforming this conversation into more feminist directions importantly consists of raising the importance of questions of justice and equity, which necessarily also entails re-thinking and re-defining the meanings and values of water. How to best do this importantly depends on the issue at stake, and always is itself an engagement with power.

Seema: I have been active in the rural women's movement in India since the 1990s. I prefer not referring to myself as a gender professional since to me the term originates in and represents a distinct donor-driven development agenda, one that tends to render 'gender' into yet another problem or deficiency that can be fixed through technical expertise. I instead belong to a generation of feminists whose work is anchored in and inspired by direct engagements and interactions with rural women, for whom water was and is a major concern. Currently, I work for SOPPECOM, an organisation working primarily on land and water issues. SOPPECOM started out as a multi-disciplinary team with engineers and social scientists working and thinking together. SOPPECOM is committed to advancing the agenda of social justice in water, with a clear focus on gender and caste issues.

In my work, I start from an identification with the lived experiences of women and from a deep understanding of their social contexts. Much of my work consists of trying to trace the likely impacts of macro policies on their lives. Hence, we collect vital information on women's work, also in agriculture – such as weeding, threshing, etc. – we document gendered patterns of land access and ownership, and highlight the importance of women's knowledges in such things as seed conservation, etc. All this helps us in understanding the irrigation conditions, such as the functioning of the irrigation system, its current users and members, and the gender and caste composition of the members and users. In a nutshell, whereas the technical team might be more interested in the technical performance of the irrigation system and the health of the canals, we would also like to understand the use of the system and its users: and this is where understanding gender relations is important.

Babette: I've been a researcher on gender and environment now working more within interdisciplinary teams on energy transitions and finding ways how this could work. I struggle with the emerging research approaches today that favour more positivist knowledge than embodied knowledge even among social science researchers and its privileging among development practitioners.

Becky: I come from an academic background in human geography, where my research on gender and environment has been heavily informed by fieldwork concerning forests and rural livelihoods. Being based in a university, where most of my time is spent teaching the next generation of gender and environment researchers, I'm interested in how apparent gaps between critical feminist theory and gender and development practice are reconciled (or not) in environmental science contexts.

Margreet: In 2014, I joined IHE-Delft, an Institute for Water Education and I am also affiliated with the University of Amsterdam. At IHE-Delft, I work with natural scientists and engineers focusing on different water questions. I am responsible for the small but rapidly growing social science corner in the Institute. In the course of my academic career (which started in 1997), I've done a lot of work on water, often on issues of equity and justice, and often through the entry point of gender with an explicit feminist commitment. I have come to appreciate how much of this work consists of creating the spaces to discuss radical and often oppositional feminist ideas, often translating them to make them palatable as well as actionable to those working in water.

I was trained as an irrigation engineer, but during my MSc studies I chose to divert from the normal engineering curriculum to take courses in Development Economics, Anthropology, Gender Studies etc. Much of this was prompted by a deep interest to understand how changes in (access to) water and changes in social relations co-constitute each other. Even while I never felt myself to be a 'real' engineer (whatever that may be), my basic training and background as an engineer was and is useful in establishing some credibility with my colleagues and wider peer group when working on water.

I am not sure whether this background has helped my credibility as a so-called gender professional as this very much depends on who is asked. As I always worked between and across disciplines, those colleagues who invest pride and derive status from belonging to one discipline sometimes question whether I 'belong' to the engineering profession or whether I am either a

> 'true' social scientist or engineer. I myself try to steer clear from such status battles, and aspire to publish in geography, development studies and social sciences journals as well as in water and environmental science journals. At the same time, I am very much aware how my own authority and influence – for instance in creating space in professional water domains for questions of gender justice – also depends on my ability to mobilise and assert disciplinary forms of authority.

Some contributors to other chapters in this book indeed highlight their 'crossovers' from their technical backgrounds to undertaking gender-focused research and interventions in different environmental and resource contexts. And like Margreet above, they too strategically use their disciplinary authority as technical specialists to legitimate attention to gender and social justice issues in highly technical working environments.

Feminists, gender experts and women engineers

Margreet and Seema are concerned that the inclusion of gender experts and expertise in water development programs stimulated by global water agreements may neutralise, instead of advance, a more political and feminist gender and water justice agenda. This is because a mandate for transformative feminist change intrinsically entails questioning accepted knowledges and related positions, identities and hierarchies. Indeed, addressing gendered inequities in the technical environment of the water sector importantly includes the critical scrutinising of established epistemological normalcies as well as the questioning of normal ways of diagnosing problems or proposing solutions. Beyond merely including women or gender, therefore, addressing questions of gender (in-)justice requires the cautious navigation of political and epistemological tensions. Feminists hardly ever have enough political leverage to simply demand or order what needs to happen, which is why they need to find ways to convince those with more power to follow their recommendations even when those are unsettling. In water, this implies being able to talk the language of engineers and understand and empathise with their pre-occupations (see for example, Udas & Zwarteveen, 2010). Doing this always carries the risk of too much identification with the engineering or technical project of efficiency or productivity, to the neglect of the goals of justice and equity.

Women engineers, for their part, may not be taken seriously by their male peers. According to Liebrand & Udas (2017), for them to succeed and belong, they have to reconcile the performances of being a 'lady engineer' with that of a 'normal' masculinised engineer, which may be asking for the irreconcilable. This brings to fore how feminist politics in water contexts also involves becoming aware of, navigating and contesting prevailing identity boundaries. This points to the difficulty for women to credibly perform as engineers, and it points to epistemic hierarchies in water professional contexts between social sciences and natural sciences engineers. Gendered professional identities directly matter here, as in water those who are generally attributed most authority are male engineers. For social scientists, let alone gender specialists or feminist scholars, it may be difficult to be seen and treated as sources of expertise.

Kunz et al. (2019) point out that gender experts as social scientists may cope with such hierarchical and political tensions by 'erasing the boundaries' between gender expertise and the positivist sciences to realise more pragmatic and technical outcomes. This may lead to de-politicising gender and water agendas to a significant extent.

Margreet: I am getting many requests to do gender related work in water projects supported by development cooperation funding. In these projects, gender often risks becoming a mere 'technical' issue, one that is totally de-politicised. Development projects place a lot of emphasis on success stories, best practices and solutions. The associated language may not be helpful in realising feminist goals, because feminism intrinsically is a process and a struggle, and not something you can achieve in a project of 4–5 years. Contemporary water projects for instance emphasise the potential of new irrigation technologies – such as drip irrigation – in helping farmers become entrepreneurs. In this storyline, farmers – either female or male – figure as individual businesspeople who are constrained in reaching their potential because of their limited access to technologies, markets or finance. The gendered version of this story line is that women face even more constraints than men, resulting for instance in prognoses of what production could be if women would have the same access to land, water, technologies and much else as men. There is much to be said against this story line, but what matters when examining it from a feminist lens is that small-scale irrigated farming – almost everywhere in the world – is a family affair in which men, women and children all collaborate, often following historically and culturally distinct ways of distributing responsibilities, rights and incomes. It therefore makes little sense to treat women or men as individual farmers, or to compare men's productivity with that of women: farming is a collective matter, and understanding it requires an understanding of the intra-household organisation of agriculture.

Discussing this with donors and those who believe in the story of the individual farmer-entrepreneur is difficult. That is the field of tension that I find myself in. And I'm trying to navigate, and I don't always find it easy.

Margreet and Seema both distance themselves from self-identifying as gender experts since they believe that being so ties them to mainstream development agendas that do not fully seek to understand women and men's lived experiences with water systems. Margreet, in particular, observes the work of gender experts as sometimes removed from the very contexts they are trying to change.

Becky: Would you say then that gender experts have engaged in a so-called 'technocratic compromise'? One of the contributors to this book said, 'Well, bring them to the field so they can see. The complexity can be demonstrated.' Maybe you have other thoughts to add to that?

Babette: That's good of you to recall, Becky. But at the same time, we're also facing the dilemma of how we can use the development space for opportunities? Or should we actually just wish them away and say, 'No, we refuse to be co-opted by the development process since it blurs our feminist and justice agenda.' And a number of development organisations and donors today support the work of gender experts and it is part of their institutional mandate to ensure they are an integral part of programmes.

Margreet: My experience is that often, development donors and implementers are under pressure to demonstrate success, something they can do by suggesting direct causal relations between interventions and positive outcomes for women or gender equity. The dynamics of power and politics that characterise feminist transformations are difficult to capture in causalities and outcomes, as these refer to processes of struggle and contestation in which winners and losers are not always easily determined, with cause–effect relations that are less straightforward and multi-dimensional. This does not mean that when I am asked to contribute to making water policies or projects more gender aware or inclusive, I refuse. I do sometimes accept because of the possibility to positively engage with planned changes or interventions. And then in the process, also ask some of the more difficult questions. So that's now becoming my strategy.

Water productivity, state reform and the water profession

Seema has contributed to knowledge on gender and water reforms in the Indian context by highlighting the missed opportunity that decentralisation provided to democratise water governance through a policy of non-engagement with social and feminist movements with alternative knowledges and cultures of water management (Kulkarni, 2011, 2016). She also demonstrates how far from reaching their inclusive ends, reforms around privatising groundwater and sanitation infrastructure have increased gender and social inequity, environmental risk, and gender-based violence (Srinivasan & Kulkarni, 2014; Kulkarni et al., 2017). Seema's continuous engagements with policy makers and NGOs have become seamless with her own research pursuits especially exploring the meanings attached to water use.

Margreet: I am always inspired by Seema's work in Maharasthra with the Irrigation Department, working in the program around increasing participation of water users, and using that to increase spaces for poor and landless women.

Seema: We are working with a techno-centric and male-centric irrigation department. In the 1990s, we were lobbying at the policy level and advocating the irrigation bureaucracy to include women in water user associations-local level institutions formed for better participation of farmers in surface irrigation systems. Gender experts were unheard of then. The irrigation bureaucrats were not prepared to talk about issues around landless women and water. They had their typical parameters and indicators to judge the success of the irrigation system's performance, i.e., how much crop per drop of water! The irrigation bureaucracy's response to our advocacy was to ask how women and landless people's participation was going to improve irrigation performance, which usually referred to how much water was served on a plot of land for crop productivity. They could not see beyond these parameters. They could see the roles of women and landless groups only insofar as these could improve irrigation performance, which was a very narrow and straightjacketed view.

Their ways of thinking did not allow them to take the broader view of people, particularly the role of marginalised people. At that point, we realised we needed to change our ways, we needed to say things differently. We needed to help them see that there's a need for a more just world. This process of change – with a lot of support from global normative discourses – helped raise awareness in the government. Since 1995, they have gradually begun to see gender in the water question.

The outcome was bringing in women and marginalised groups under the policy [legal reform] and decision-making bodies. This meant that they do recognise their roles. I know we need to go beyond representation, but what's happened is a really big step forward. Yes, representation is necessary, but that is not sufficient. There have to be commitments that representation should lead to broader outcomes – just water practices. There was a lot of learning among communities too. We still have a long way to go helping bureaucracies recognise the politics of gender and water. I think it unfortunately has not moved beyond representation and quotas. Within irrigation water user associations, there is representation of women. For domestic water, there would be village water and sanitation committees. These are local level formations. State-regulated structures are federated with water distribution from dams and canals to the irrigation project level. Irrigation would have project boundaries not administrative boundaries. So, there is representation of these groups in these committees. But in the profile and composition of the bureaucracy itself and in the technical jobs and the hierarchies, you will not see women or marginalised groups represented. It is only because of the quota system for caste that we see some appointments of marginal groups. There is an absence of women at the top positions in water bureaucracies.

The discourse of 'water for productivity' is also changing. Over the decades, organisations like ours have changed the discourse such that it is no longer politically correct to make that distinction between productive and non-productive uses of water, although the mindset on productivity is still very strong. There is a greater recognition now of the multiple uses of water. A very generic water policy document will state that, but there's little buy in. But when you look at the programmatic documents and actual allocations, you again see that they're largely channeling water for paying sectors. In that sense, industry gets the priority. It still is predominant, but whenever there is pressure, it is no longer easy for bureaucracies to get away with that kind of talk [the productivity argument]. Shifting the mindset is an ongoing challenge, nevertheless.

On the subject of water productivity, Margreet, for her part, wonders why this is a prevailing mindset among water professionals.

Margreet: What I find very odd, working on water as a feminist, is that the predominant language used by my water colleagues is a language of productivity and efficiency: it's about the dollar per drop of water. As feminists, we use the language of justice, of equity, one that allows recognising the sharing and care for water. My water accounting colleagues use remote sensing data water maps that are in theory accessible and legible to many, and thus could be mobilised in support of improving water democracy in theory. But if the maps only show water productivity, what kind of water futures do they support? So sometimes I ask my colleagues, can we not have 'water happiness or wellbeing maps' instead of only making water productivity maps? We're only starting these discussions. It requires so much patience to actually sit together and understand each other.

Living with technical and social incongruities in water management

There is increasing concern over a prevailing type of rational environmental governance where epistemic authority is vested on mathematical and statistical abstractions in the form of statistical indicators, modelling, imaging technologies and risk assessment exercises (Wang, 2015). This is an effort to render 'unruly' nature (e.g., water, climate and forests) intelligible, ascribe it with more certainty, and therefore depict it as more amenable to control, prediction and management. This technical 'evidence-based' approach has been popularly used to premise key policy decisions, which, however, deflects attention from the drivers of inequality and disadvantage (Eyben, 2013; Bacchi & Goodwin, 2016; Eriksen et al., 2015; Taylor, 2014). In contrast, Margreet and Seema's corpus of work has been dedicated to disrupting such efforts by showing how prevailing ways of measuring and accounting for water only make some characteristics or capacities of water visible – such as those aligned with colonial discourses of avoiding waste and improving efficiencies – at the neglect and expense of injustice and exclusion issues. In addition, scientific conventions of measuring water treat it as if it can be disconnected from social relations and histories – relations of labour and property, most notably – which, as a result, also make it impossible to see water as deeply connected to and constitutive of societal orders. Questions of gender inequality, patriarchy, and injustice therefore remain very difficult for water scientists and professionals to recognise and tackle.

Prevailing scientific and professional ways of knowing and conceptualising water are themselves the legacy of particular colonial and patriarchal structures. Margreet has scrutinised masculinity in the irrigation and engineering profession by unpacking its cultural, material and performative elements and highlighting processes of masculinity's normalisation and naturalisation; citing masculinity as a product of the diffusion of colonial and professional water knowledge that remains persistent; showing that to be an engineer is to embody masculine qualities such as scientific rationality and positivism, technological prowess and physical hardship; that the engineering sciences have an almost stubborn propensity to bracket out knowledge on gender and social relations; and how water rights have been conventionally narrowly defined as male-controlled due to fewer possibilities for women to own land and other vital resources that mediate water rights (Alda-Vidal et al., 2017; Delgado & Zwarteveen, 2017; Zwarteveen; Zwarteveen, 2008, 2010, 2011, 2017; Zwarteveen & Rap, 2017; Zwarteveen et al., 2012). These ideas rally around a key theme in this book that suggests gender, power, knowledge and professions are indeed co-constituted.

Becky: One of our contributors said that they found it easier in some ways to engage with some of the technical people who had a background in agricultural systems or in rapid rural appraisals as they'd done field-based work, which necessarily meant working with people. I don't know whether the same could be said in the water sector or if there's something specific about water and its connection to engineering that means it's sometimes quite disconnected from the field or at least from voices in the field. Does this make it difficult to negotiate and find a common language around social issues?

Margreet: Yes, I think that's a very good point, Becky, because I remember years ago when I was working with the International Water Management Institute (IWMI) where at that moment we had the conversation across different international agricultural research centers about how to do gender, or how to make more space for gender. And the consensus seemed to be that participation was the entry point: participatory approaches to development. And I thought, well, water participation seems nice. It's a good entry point. But over the years, I have come to think that often questions of participation and decision making do not touch on wider allocation or distribution questions. So, I also felt we are missing out some of the more fundamental questions if we approach it from the angle of participation.

And you're also right, I think that engineers are not wired to diagnose problems, but their minds are wired to propose and design solutions. So that bringing them back to consider what problem a particular solution will help solve is a difficult step to make. Because for them, it feels like going backward to what the problem is. For them, it's all a matter of 'let's get on with it!' That's a comment that I often hear. Because when you want to do gender, actually you need to make that step back and rethink, 'Hey, but what is the problem? Can we rethink it?'

Babette: Or maybe – correct me if I'm wrong – because engineers actually define the problem in very technical terms, such as: 'Oh, there has to be a bridge somewhere or some infrastructure that's needed.'

Margreet: Yes. But so that is not even defining a problem, it's defining a solution. It's proposing a solution to a problem that is supposed to be already diagnosed or identified or known.

In her own earlier writing, Margreet says that 'unwillingness or persistent biases of individual water analysts, the difficulty to see women and gender in water is linked to a particular epistemic tradition in knowing water that is deeply inhospitable to the analysis of social relations and gender. An important conclusion is that "thinking" (and acting on) gender in water management also and importantly requires active efforts to change normal ways of knowing in water' (Zwarteveen, 2010: 75).

Babette: I ask myself constantly why it is difficult for engineers and modelers to link gender issues with their work. This was especially true when I was teaching in a predominantly engineering university. The only option was for us gender researchers to work separately from the engineers. It may be because their worldview is different, and perhaps their methods too. We existed in ontological and epistemological silos. And that created an uneasy peace. Now working in a different organisation that is open to inter and transdisciplinary work, I am pleased to find water modelers who are excited to integrate gender into their work. However, they are limited by the scarcity of gender-disaggregated data in water systems, so their recourse is to confine gender concerns to including women in stakeholder engagement. I wonder whether it is just a matter of acquiring gender-disaggregated data – which in their view achieves gender integration – or whether it is a matter of different types of knowledge. I must confess that I need to know and understand modelling more to know how and where the useful entry points for gender are. Or maybe I suspect there are no entry points, especially since their approach is disaggregation, that is, to count women and men; whereas our research approach is to understand embodied experiences.

Margreet: In modelling, all kinds of assumptions are made about which descriptors to use for representing a reality. In water, a very basic assumption that often goes into modelling is a normal natural science assumption: that it's possible to somehow compare waters across many different places. This requires making abstractions about water that always exclude social dimensions. Once the model is used for decision making – it will have to be re-contextualised – and so there, gender can matter, but it's often not included in the re-contextualisation.

Some of our conversations (on epistemology) have led to fights. They (modellers) are working from a positivist perspective. As a social scientist, we use more constructivist, non-modernity approaches. I realise that it's almost impossible to bridge these two different ways of knowing. But I think it is possible to start the dialogue by asking whether there are other ways of representing water (other than for productivity), or to model for different water logics. I felt it was a bit easier to start with the question of logics than with the question of knowledge, because with the knowledge question the modelers felt that I was telling them that they were not being objective. In a way, that's how they understood it, but I told them that it was not meant as a criticism. What I wanted to say to them was that none of us are objective, at least not in the conventional or popular use of the term. But that was just one step too far for them.

Babette: Interesting. And I think this brings us to the next point, which is really about different and multiple knowledges in relation to studying and managing water. Are there better ways for the technical and the social to co-exist? We often invoke the positive outcomes of interdisciplinarity or even transdisciplinarity and the need to work with cognate fields. Maybe it's better to think not in terms of integrating knowledges, but really 'juxtaposing' knowledges or a kind of 'triangulation'? That's a thought I want to sort of just throw in.

Researchers conventionally use triangulation as a methodological approach to ascertain whether the data collected through one method matches the data collected through another method. It is mainly used to ensure that the data collected is reliable. This is done by collecting data using diverse methods, recognising the internal validity of each of the data sets collected respectively through such methods, but at the same time interrogating the 'silences' between the data collected, and finally recognising the limits of each method (Nightingale, 2003; Hesse-Biber, 2012; Harris, 2016; Nightingale, 2016). These silences allow for generating further research questions that otherwise may have been omitted or sidelined in the original research design but can potentially shed light on unexplored realities. The approach also underscores the point that feminists have been making all along: that all knowledge is partial (Haraway, 1988). It also presupposes that the research can be undertaken by researchers from diverse research and disciplinary backgrounds. Because of this, researchers have come to mine its potentials for enabling cross-learning and knowledge co-production.

In the water sector, Mukhtarov and Gerlak (2014) have come to recognise the importance of epistemic pluralism in addressing the challenges of water management that are far too complex for the prescriptive nature of integrated water resource management. Their work draws attention to the co-existence and versatility of multiple knowledges that go beyond the idea of triangulation as a methodological approach and recognise the need to capture and interrogate diverse knowledges in water contexts.

Margreet: I like the idea of triangulation a lot. For a long time, the MSc program on water management at the IHE was partly implicitly based on the idea of integrated water resource management. So, the idea of integration became very central to the whole program. Triangulation may help re-think what integration implies and does. The idea of integration actually suggests that it's possible and desirable to bring all different knowledges together into one big whole, a *uni*-verse. This may not be the most fertile idea, as different knowledges and ways of knowing cannot always be easily added up or integrated:

there are resonances, but also frictions, clashes and indeed silences. So, what we are now experimenting with, is rather than integration, use some idea of plurality linked to controversies. Rather than making integration and harmonisation the guiding ideas and implicit ideals, our attempt is to direct attention to plurality, difference and contestation. By acknowledging differences of interest, differences in diagnosing problems, different solutions it becomes possible to accept that perhaps there is not one world, a *uni*-verse, but there are many possible worlds, a *pluri*-verse. I don't know whether that is what you mean with triangulation, but if it means creating room for different ways of knowing and different versions of reality to co-exist I am all in favour.

Babette: But what seems to be quite mainstream up to now is the integrated water management approach. And do you think there are opportunities perhaps to realise an altogether new agenda for water management, given this need to juxtapose, differentiate, rather than integrate?

Margreet: I would hope that in triangulation, we would also be learning to disagree and accept that there are differences that are sometimes very difficult to reconcile. And of course, compromises need to be made if you need to arrive at a practical solution. But there are always differences of interests, contestations and conflicts that the word triangulation may better allow to acknowledge, whereas the word integration entails an active invitation to forget about those by focusing on consensus or collaboration. Resolving irreconcilable elements is always a practical and political concern, as much (or more) as it is a scientific one.

Becky: One of the issues we explored with other contributors was the politics of knowledge. It is actually also about who is setting the terms of the dialogue because some forms of knowledge are dominant in those exchanges. It's like gender. People are invited in to do the 'gender work'. But, actually, they're not really driving the wider agenda or even dialoguing on the wider agenda, or indeed setting the initial terms of the debate. I suspect that quite a lot of them are invited but not on their own terms.

Seema: Yes. It is really only worthwhile when there is an opportunity to dialogue, but obviously not where one or the other is sort of dominating the terms of that dialogue. So, if it is an open process of dialogue between the technical and gender professionals, I think it's worthwhile to have.

Margreet: I have come across many water projects that have hired people to do 'gender' in their projects. Sometimes their most important qualification for being hired was that they were women. What I have seen happening and what still happens is that these gender professionals are put in a position where they continuously have to explain and defend what they do; why attention to women and gender is important. One dangerous strategy that some of them resort to is to approach gender equity as primarily a moral or ethical position, something that belongs to either individual integrity or that is an indicator of 'development', with an idealised and non-existing gender equality in western countries figuring as the implicit norm that others should follow or adhere to. I have seen very awkward situations where white, western gender professionals preach to for instance Nepali or Indian irrigation engineers on how they should deal with women or gender in the water projects they are professionally involved in. Awkward not just because of how these gender awareness trainings make often wrong assumptions about what these engineers think about and how they deal with questions of gender, but also because of how they misrecognise the few real practical possibilities that engineers have to address questions of gender equity in their work. Gender equity of course always is an ethical or moral question, but it is also a practical and political one and one that always transcends water. Therefore, what gender equity is or should be cannot be formulated on the basis of abstract ideals but needs to be anchored in grounded gendered water experiences and histories. Engineers working in the field always do have knowledge about gender relations, and this knowledge should be the basis for engaging in discussions about what they can or should do. I think that gender is so complex, and it's only by allowing people – both irrigators and water users as well as engineers – to speak for themselves that one can start having meaningful conversations. I therefore also think one is in a much better position to do gender work with technical water professionals if one understands, acknowledges and respects what they are doing.

Babette: Would gender experts or feminists on the ground need to know something about engineering? And if they have social science backgrounds, would it be necessary for them to have skills to deal with the technical?

Margreet: Yes, I think so. You don't have to be an engineer, but you should have some respect for the work that the engineers are doing. I think Seema is always saying that there seem to be very few good social scientists interested in gender and water. At the same time, if the context in which you work is a very technical one like the one I found myself in the International Water Management Institute in the 1990s, you are continuously pressured to translate between different languages. I don't know whether you need a technical background but, at least, you need to learn how to do this translation. And you also need to start understanding what gets lost in the translation.

Seema: I agree generally with Margreet but I also think that the reverse is true. I mean, you also need technical people with an understanding of the social questions, I would say, a sensitivity to it, and also the need to see the interface between technology and society, which, unfortunately, is very, very lacking in the context that I am working in actually. And I also feel that the burden for people like us who are talking about social justice, differences, and discrimination, to understand the technicalities of water is much more, because we have to know the sector extremely well to be able to see and argue in a very firm manner, to show that there is discrimination, and in what possible ways we could address it. So, I think the burden is much more on people like us (social scientists/practitioners). It is possible for the water agenda to move forward without having to address any of these social or equity concerns. For example, transboundary issues, or questions around agriculture versus industry, contestations between various uses of water. I mean, even if you

> don't say a word about gender and social relations, that agenda still moves ahead, pretty much, and very easily. But the same is not true if I'm only going to say 'gender, gender, gender', it will completely be in a vacuum, and I'm never going to be heard. So I know that I have to be very thorough with my understanding and knowledge about transboundary water issues or contestations in this terrain, if I want to be heard on questions of social difference. We have to reach the stage where people who are looking at the technical aspects, or I would say even people from social sciences but who don't really look at social discrimination and social justice have to be sensitive, and educated enough to say, 'No, the agenda cannot move forward if we have not understood the question of social difference'. A two-way exchange is very, very strongly needed. Because I have seen that we have just been silenced out in discussions around, for example, drip irrigation. As if there is no social agenda that exists, it's just a matter of putting those systems in place. So the burden was much more on advocates like us who then had to make sure that we understood drip irrigation, its technology and its implications on questions of allocation and access.

To conclude

Our analysis suggests that the incentive structure and professional culture that prevails in most public water bureaucracies stands in the way of achieving the goals of women's inclusion or empowerment. Public water professionals – the large majority of whom continue to be engineers and men – are formally rewarded and informally appreciated for the construction work that they do, or for the number of the projects they are involved in. Professional engineering cultures tend to naturalise or normalise associations between professional performance and masculinity, making it difficult for women to become respected members of the professional water community. The gender experts in water development projects often lack the political clout to change the terms of dialogue with their technical colleagues, as their work is often considered marginal to the main task of achieving water productivity.

We end this discussion with the view that there are different and hierarchical ways of knowing in the water sector, where epistemic privilege mostly still lies in creating physical and institutional conditions for water productivity to serve market-driven ends, rather than for purposively democratic and gender-just water distribution. Integrated water resource management has served as a popular vehicle to merge these types of knowledge and tie them to a strong participatory ethic. Years of observing the application of these approaches through the creation of water user and other water subsidiary groups have led us to conclude that just and re-distributive aspects of water continue unaddressed. Indeed, there may be a need to re-think the widely used view of 'integration' and shift instead to recognising the constancy of irreconcilable difference and contestation around water as a resource – or even better, seek more creative and promising ways of co-producing knowledge that recognises that there is more than one way of knowing.

References

Alda-Vidal, C., Rusca, M., Zwarteveen, M., Schwartz, K., & Pouw, N. (2017). Occupational genders and gendered occupations: The case of water provisioning in Maputo, Mozambique. *Gender, Place & Culture*, *0*(0), 1–17. https://doi.org/10.1080/0966369X.2017.1339019

Bacchi, C. L., & Goodwin, S. (2016). *Poststructural Policy Analysis: A Guide to Practice*. Palgrave Macmillan.

Delgado, J. R. V. & Zwarteveen, M. (2017). Queering engineers? Using history to re-think the associations between masculinity and irrigation engineering in Peru. *Engineering Studies*, 22.

Eriksen, S. H., Nightingale, A. J., & Eakin, H. (2015). Reframing adaptation: The political nature of climate change adaptation. *Global Environmental Change*, *35*, 523–533. https://doi.org/10.1016/j.gloenvcha.2015.09.014

Eyben, R. (2013). *Uncovering the Politics of 'Evidence'and 'Results': A Framing Paper for Development Practitioners*. Prepared for the Politics of Evidence Conference in Brighton, UK.

Gilmartin, D. (2003). Water and waste: Nature, productivity and colonialism in the Indus Basin. *Economic and Political Weekly*, *38*(48), 5057–5065.

Haraway, D. (1988). Situated knowledges: The science question in feminism and the privilege of partial perspective. *Feminist Studies*, *14*(3), 579–599.

Harris, L. M. (2016). Intersections of gender and water: Comparative approaches to everyday gendered negotiations of water access in underserved areas of Accra, Ghana and Cape Town, South Africa. *Journal of Gender Studies*, 1–22. https://doi.org/10.1080/09589236.2016.1150819

Hesse-Biber, S. (2012). Feminist approaches to triangulation: Uncovering subjugated knowledge and fostering social change in mixed methods research. *Journal of Mixed Methods Research*, *6*(2), 137–146. https://doi.org/10.1177/1558689812437184

Kulkarni, S. (2011). Women and decentralised water governance: Issues, challenges and the way forward. *Economic and Political Weekly*, *46*(18), 10.

Kulkarni, S. (2016). Gender and water in India: A review. In V. Narain & A. Narayanamoorthly (Eds.), *Indian Water Policy at the Crossroads: Resources, Technology and Reforms: Vol. Global Issues in Water Policy* (pp. 73–91). Springer International Publishing.

Kulkarni, S., O'Reilly, K., & Bhat, S. (2017). No relief: Lived experiences of inadequate sanitation access of poor urban women in India. *Gender & Development*, *25*(2), 167–183. https://doi.org/10.1080/13552074.2017.1331531

Kunz, R., Prügl, E., & Thompson, H. (2019). Gender expertise in global governance: Contesting the boundaries of a field. *European Journal of Politics and Gender*, *2*(1), 23–40. https://doi.org/10.1332/251510819X15471289106112

Liebrand, J. & Udas, P. B. (2017). Becoming an engineer or a lady engineer: Exploring professional performance and masculinity in Nepal's Department of Irrigation. *Engineering Studies*, *9*(2), 120–139. https://doi.org/10.1080/19378629.2017.1345915

Mukhtarov, F., & Gerlak, A. K. (2014). Epistemic forms of integrated water resources management: Towards knowledge versatility. *Policy Sciences*, *47*(2), 101–120. https://doi.org/10.1007/s11077-013-9193-y

Nightingale, A. J. (2003). A feminist in the forest: Situated knowledges and mixing methods in natural resource management. *ACME: An International Journal for Critical Geographies*, 77–90.

Nightingale, A. J. (2016). Adaptive scholarship and situated knowledges? Hybrid methodologies and plural epistemologies in climate change adaptation research. *Area*, *48*(1), 41–47. https://doi.org/10.1111/area.12195

Srinivasan, V. & Kulkarni, S. (2014). Examining the emerging role of groundwater in water inequity in India. *Water International*, *39*(2), 172–186. https://doi.org/10.1080/02508060.2014.890998

Taylor, M. (2014). *The Political Ecology of Climate Change Adaptation: Livelihoods, Agrarian Change and the Conflicts of Development*. Routledge.

Udas, P. B. & Zwarteveen, M. Z. (2010). Can water professionals meet gender goals? A case study of the Department of Irrigation in Nepal. *Gender & Development*, *18*(1), 87–97. https://doi.org/10.1080/13552071003600075

Wang, T. (2015). Green governmentality. In R. L. Bryant (Ed.), *The International Handbook of Political Ecology* (pp. 318–331). Edward Elgar Publishing.

Zwarteveen, M. (2008). Men, masculinities and water powers in irrigation. *Water Alternatives*, *1*(1).

Zwarteveen, M. (2010). A masculine water world: The politics of gender and identity in irrigation expert thinking. In R. Boelens, D. H. Getches & J. A. Guevara Gil (Eds.), *Out of the Mainstream: Water Rights, Politics and Identity*. Earthscan.

Zwarteveen, M. (2011). *Questioning Masculinities in Water*, *18*, 10.

Zwarteveen, M. (2017). Hydrocracies, engineers and power: Questioning masculinities in water. *Engineering Studies*, *9*(2), 78–94. https://doi.org/10.1080/19378629.2017.1358730

Zwarteveen, M., Ahmed, S., & Gautam, S. R. (Eds.). (2012). *Diverting the Flow: Gender Equity and Water in South Asia*. Zubaan.

Zwarteveen, M., & Rap, E. (2017). Guest Editor's Introduction: Engineering masculinities in water governance. *Engineering Studies*, *9*(2), 75–77. https://doi.org/10.1080/19378629.2017.1363019

7 Embodied engagement with gender and agrobiodiversity

Leveraging transformative moments in multidisciplinary teams

By Rebecca Elmhirst and Bernadette P. Resurrección
In conversation with Marlène Elias

Introduction

What happens when a leading international research centre that provides technical knowledge on agricultural and forest biodiversity to strengthen food security and ecosystem health is required to bring a gender perspective into its work? As part of a strategic initiative, Gender Specialist Marlène Elias was hired to take on this mantle in Bioversity International, one of the 15 CGIAR research centres (formerly the Consultative Group on International Agricultural Research). Bioversity International (hereafter, BI) was initially established as the International Board for Plant Genetic Resources and subsequently operated under the name of International Plant Genetic Resources Institute. Its early work focused on the emergency conservation of crop genetic resources in gene banks. The organisation's focus has changed over time, and at the time of our discussion with Marlène, BI's mission was to deliver scientific evidence, management practices and policy options to use and safeguard agricultural and forest biodiversity to attain sustainable global food and nutrition security. Since 2020, BI is in a formal alliance with CIAT (the Center for Tropical Agriculture), another CGIAR centre. The alliance of Bioversity International and CIAT aims to 'deliver research-based solutions that harness agricultural biodiversity and sustainably transform food systems to improve people's lives'. Bioversity International is one of the CGIAR centres contributing to the CGIAR Research Program on Forests, Trees and Agroforestry (FTA), which is the world's largest research for development program to enhance the role of forests, trees and agroforestry in sustainable development and food security, and to address climate change. Within her remit as Gender Specialist in the realm of Conservation and Management of Forest Genetic Resources, Marlène's work has involved implementing FTA's cross-centre Gender Strategy, launched in 2013, and more generally, integrating gender into BI's activities.

Gender research has been a part of the CGIAR's work for many years, as a substantial grey literature testifies (van der Burg, 2019). However, the relatively recent appointment of gender specialists (including Marlène) has been part of a more centralised, rigorous and measurable integration of gender issues within CGIAR around the time it launched its CGIAR Research Programs (CRPs), including FTA, in 2011. Each CRP was required to develop a Gender Strategy and provide adequate resources towards its implementation to develop gender disaggregated research, form partnerships and alliances with external gender experts at various levels, and increase knowledge sharing and learning across and beyond the network of CGIAR research centres (CIFOR, 2013; FTA, 2020). The reach and effects of this recent effort to mainstream gender in agricultural and forest research for development has attracted interest both within and outside the CGIAR (Arora-Jonsson, 2014; Arora-Jonsson & Sijapati Basnett 2018; Mukhopadhyay & Prügl 2019; van der Burg 2019). These discussions have provided a useful reflection on the underlying norms that shape what counts as knowledge in environmental organisations such as those making up the CGIAR, and on the artefacts – the tools and texts – that emerge as 'gender' is brought into agriculture and forestry research for development.

In this chapter, we bring another strand into the debates represented in this emerging literature, and offer a three-way discussion that reflects on the embodied, everyday and practical challenges of working as a gender specialist in an environmental research for development organisation, and the small pathways that offer the potential to realise commitments to transformative change. Our discussions are framed through feminist political ecology and the politics of situated knowledges, and with recent discussions around pedagogy that have been taking place in parallel feminist development studies contexts (Oberhauser, 2019; Harcourt, 2019). Part of the discussion in this chapter explores how the tenets of critical pedagogy (Freire, 2000; hooks, 1994), and in particular, experiential learning (Kolb, 1984), resonate within the challenges of bringing gender expertise into multi-disciplinary teams in largely biophysical research-for-development settings.

The chapter is organised around three related thematic areas that emerged initially from Marlène's reflections on her work, and that we explored in more detail in our subsequent discussions: (i) how pathways for embedding gender knowledge are enabled (or closed down) by the specific institutional context into which they are introduced; (ii) embodied and field-based experiential approaches for incorporating gender expertise into multi-disciplinary teams in technical settings; and (iii) the challenges of enacting such approaches at scale within the logics of technocratic project requirements. As our discussions show, the kinds of transformations required meant going beyond calls to develop a common scientific language with a mainstreamed technical application of gender expertise, and instead taking seriously other ways of working with gender in 'agriculture for development' (A4D) that develop knowledge collaboratively in ways that are meaningful for scientists, as well as for people and their communities. We begin below with a discussion of how the institutional context at BI has been important in shaping how 'gender knowledge' enters and is engaged with, within an applied science team.

Bringing gender to Bioversity International

A question we raise in the introduction to this book is whether the epistemological interface between gender knowledge and environmental science raises particular questions and issues for the 'gender expert' and the potential for meaningful (feminist) change as gender work involves engagement in techno-scientific and managerialist fields with logics at odds with the feminist social science on which much gender expertise is based. At Bioversity International, that epistemological interface has taken a particular form, shaped largely by the organisation's history and approach, and its relationship with other centres within the CGIAR.

Marlène suggests that the systems approach taken by BI to the understanding of environmental management and agriculture has offered relatively fertile ground for considering people, and with that, gender questions.

Marlène: Bioversity looks at biodiversity in farming systems and the ways that cultivated and uncultivated species come together to make up the agro-ecosystem. We are tasked with protecting biodiversity but with a view to it being used by people: managing biodiversity for its use in improving lives and healthy ecosystems, taking a systems approach. With this approach, it is easier to see where people fit in, to put people into the picture.

A holistic approach that eschews separating people from nature conceptually and in practical action has provided an institutional context in which it has been possible to open up conversations about gender equality, as many scientists within BI were already engaged in farming systems, farmer-first and participatory research approaches, some having been instrumental in this shift (Conway 1985; Chambers et al. 1989; Friis-Hansen and Sthapit 2000; see also Chapter 2). Early on, such approaches had attempted to foster new types of partnerships between farmers and technical and social scientists, and in so doing, move some way towards a more nuanced approach to what knowledge is, or which knowledge should be considered more real or valuable.

Marlène: There has been quite a bit of farming systems and participatory research originating from scientists within Bioversity International. I'm lucky to be sitting here – my colleagues are open minded and many are already experienced with working with people. I was hired to bring a gender perspective into the forestry team. I soon realised this meant bringing in a social scientific perspective and humans into forests – it was a much broader mission than 'gender'. Within that, of course, a focus on gender relations.

Marlène's arrival at Bioversity International coincided with – and was enabled by – the creation of CGIAR Research Programs (CRPs) that cut across different CGIAR centres. These were established to bring together complementary skills and knowledge that could be applied for resolving 'wicked' problems around healthy ecosystems, food security, climate change, and more, for which a systemic approach was required across the research centres internationally. Within this, Bioversity International joined a CRP focusing on Forests, Trees and Agroforestry (FTA), led by the Center for International Forestry Research (CIFOR, see Chapter 9) and involving World Agroforestry (ICRAF, now merged with CIFOR) and CIAT. Bioversity's involvement in FTA emerges as a second factor that Marlène cites as having made a difference in how bringing gender knowledge into a largely biophysical and technical setting has played out.

Marlène: The first round of these research programs was initiated to collaborate across CGIAR centres as a lot of work was being done in parallel while there were complementarities that could be brought together. At the same time, there was a push to integrate gender across these research programs, which were funded by the CGIAR Trust Fund [i.e. from donor countries, including the UK and USA]. Several donors were pushing for greater gender integration in the CRPs; the large research programs which encompassed most of the work the CGIAR was doing.

In FTA the director was very receptive, and FTA was one of the more proactive research programs in terms of putting money aside for developing and implementing a gender strategy. There is also a particularly strong social science capacity within CIFOR [Center for International Forestry Research], the lead center of FTA.

Working with other gender experts within FTA has enabled something akin to a transnational community of practice to emerge (Wenger, 1999). Marlène's efforts to embed gender within research projects at Bioversity International were also supported by collaborations with gender scientists performing similar work in other CRPs and CGIAR centers. Describing this process from the vantage point of CIFOR, the lead partner in FTA, Arora-Jonsson and Sijapati Basnett (2018) characterise the CGIAR gender network as representing multiple 'agents of change' including donors, top management, researchers, gender experts (such as Marlène) and hard incentives. Taken together, these have simultaneously diffused gender theories, approaches and training and enhanced the institutional spaces in which these could be fruitfully implemented within research projects, although as Marlène notes, there is still plenty of work to do.

In their foreword to a recent collection on the politics of feminist knowledge transfer and gender training (Bustelo et al., 2016), Marx Feree and Verloo argue that more effective change strategies are likely where there is better awareness of how knowledge is situationally conditioned (2016: xi). In the case of Bioversity International, the situational conditioning of gender equity work has come about through the entry point that earlier work in farming systems and participatory research offered, as this had already brought aspects of participatory research into technical fields and an acceptance of a more pluralistic approach to knowledge (including that of communities in which BI research teams were working). It has also been shaped by the wider institutional support that came about through a convergence of donor requirements and the right alignment of expertise that was able to influence research agendas (i.e. senior scientists at CIFOR in particular with expertise in social science and on gender). However, whilst donor checklists and tools may have made the initial push at BI, Marlène stresses that they cannot, by themselves, transform. As the next section explores, an implicit understanding of her institutional context and her efforts in working alongside her technical science colleagues has underpinned Marlène's approach to bringing gender expertise (and with this, the goal of gender equity) into the multidisciplinary research teams that are responsible for delivering Bioversity International's mission.

Decentred authority in bringing gender expertise to multidisciplinary teams

As the first person to be hired at Bioversity International with gender in her terms of reference, Marlène's appointment was very much oriented towards the support function of gender mainstreaming, and her remit involved supporting work on the conservation and management of forest genetic resources across various research projects so that this work could be gender-responsive. Key to this has been the requirement to work in multidisciplinary teams, and indeed having the fluency in the languages and epistemologies of other disciplines (forest management science, for example) is one of the qualities that earned Marlène her post at BI. Central to her work is engagement with those who are close to the field, involved in projects around forest conservation and management, including forest-based livelihoods and market integration. Initially, locating where gender might fit within the scheme of research projects involved identifying entry points within existing technical research projects to make them more gender responsive.

Marlène: Part of my terms of reference was to keep up to date with gender theories and approaches and translate those into resource materials and training possibilities. We have done this in different ways. When I began, we tried to get people talking about gender, with a number of workshops where we developed some of these basic 'what is gender' activities; in other words, participatory interactive thinking around what gender was, trying to understand and meet people where they were at that point, as researchers with limited exposure to what gender was or how it fitted in their work. We also developed a number of tools, such as on strategies for doing gender responsive data collection or communicating findings in gender responsive ways, and briefs on how gender relates to different aspects of research.

One of the approaches that Marlène has adopted was based on mentoring through funded fellowships, where researchers in five countries where BI works were recruited to work on gender equity within existing project teams and supported pedagogically.

Marlène: Probably the most effective has been a fellowship program where we had five professionals from different disciplines in different countries join project teams. Three were young and two were older researchers who had not worked on gender before. They received mentorship, used participatory methods in the field and tried to influence their project teams, and learned together through social learning as they moved through the program. I don't know how much they influenced the larger projects in which they were embedded, but they did generate research products that helped to put gender on the map at BI. It made other researchers feel like something was happening and enabled them to see more connections about how gender relates to the work they were doing.

The approach Marlène describes in the work she has done to bring gender knowledge into BI – social learning, enabling people to see connections – carries echoes of Brazilian popular educator Paulo Freire's critical pedagogy (Freire, 2000), which had underpinned much of the early epistemological shift in the 1980s to participatory modes of development training and practice that aspired to be reflexive, inclusive and non-hierarchical (Chambers, 1993, see also Prügl, 2016). In contrast to a 'banking' model of knowledge transfer, the emphasis is on learning as a process that is embodied, reflexive and experiential. In our discussions, Marlène's reflections illustrate the ways in which she has implicitly decentred her authority as a 'gender expert' in her approach to building gender work within BI. The moments she identifies as successful are where she has avoided being heavily didactic within multi-disciplinary research teams and instead encouraged an experiential and affective mode of learning based on discovery. As she notes, this began early in her work at BI.

Marlène: One of my colleagues who had been on my interview panel said that 'part of what made us select you was firstly that you had a forestry background, that you could speak our language, but secondly that we found you had a softer approach as to how you wanted to bring us somewhere, accompany us in our thinking on how to bring gender in'. That was very interesting to me. I didn't do that deliberately, but that has very much been how I've tried to work. It has been a privilege to work with a small team; it has been a conversation, a dialogue. You are not going to take the conversation in the same way with different people. The interpersonal approach is very much to the point. What resonates with one colleague is not the same as with another. It is not about trying to convince people. In Québec, we say: '*L'essayer c'est l'adopter*', which means that if you try something good, it will speak for itself – and you'll want to do it again, to adopt it (take it up). I feel excited about something others can discover in their ways and in their own time. What validates us as professionals? Those moments when I feel someone is seeing something in a new way than before. I remember the first time I read Edward Said's *Orientalism* [Said 1978]. From one moment to the next, my world changed. I remember very vividly thinking that before, I was blind; now my world has got a little bigger from reading this.

In outlining her own learning process and personal development, what emerges in Marlène's reflections is a sense that she has implicitly adopted a mode of working aligned with approaches being explored in recent feminist discussions that bring an explicitly intersectional feminist pedagogy into gender training in development in which the gender expert – effectively the pedagogue in this context – decentres their authority and learns with those in their team (Harcourt, 2019; Ferguson, 2019; see also hooks 1994). Such discussions argue that rather than persuading, gender training should emphasise process rather than outcomes (Mukhopadhyay & Wong, 2007; Davids & van Eerdewijk, 2016), should point to the need to validate the personal experience of learners (Prügl, 2016; Harcourt, 2019) and encourage a commitment to social justice, critical thinking and open-mindedness (Ferguson, 2019). This contrasts with the more commonly practised top-down approach to gender training, where the emphasis is on 'frameworks and checklists that can be easily applied' and the simplification of 'complex theories and context-specific analysis into easy and digestible terms' (Davids & van Eerdewijk, 2016: 87). Technocratic aspects of gender mainstreaming do form part of Marlène's work: 'Some might be checklists that offer an approach and guidance for working but are not necessarily inspiring'. However, she notes that: 'For any real change, there must be a long-term process, for people to change and see how things come together. A lot of that is experiential, you have to play around with it, you have to feel it, you have to get in there.'

We asked how, in the process of interactions and dialogues within her teams, Marlène deals with differences in terms of peoples' backgrounds, positionality (in terms of gender, race, cultural difference) and, given the premise with which we opened our discussions, difference in terms of biophysical and technical academic backgrounds. Marlène has noted that knowledge hierarchies between technical scientific and social science knowledges within BI were perhaps less entrenched than in other environmental research-for-development contexts. But how does this more pedagogic approach play out within BI, which is comprised of a range of expert knowledges and technical-scientific professional identities? Marlène's reflections suggest this requires more than validating the *personal* experience of others (Prügl, 2016; Harcourt, 2019). A multi-disciplinary context linking bio-physical and social science also means 'engaging with others whose "truth commitments" are structured around different premises' (Marx Feree & Verloo, 2016: x). This involves working with colleagues' existing knowledge base and experiences, recognising the value of this knowledge and then building upon this to help them learn (Freire, 2000; hooks, 1994).

In contexts where it is sometimes difficult for social science, feminist and qualitative forms of knowledge to be heard, this also means encouraging an environment in which people are respectful of each other's knowledge and expertise, and recognising where it is more appropriate to source the right skills for the right projects, rather than undertake gender training that leads people to feel they are gender experts but without the depth of study or engagement that is required for this to be successful. The 'softer ground' for this kind of approach that was described in the earlier section, might be one factor that allows for a flourishing of multi-disciplinary work and the valuing of gender expertise without the obstacles that accompany unhelpful knowledge hierarchies.

Marlène: You want people to be able to do gender-responsive work. For example, I encourage people to think about who they are asking and what they are asking [in relation to collecting social data]; to think more deeply. But doing *gender research*, trying to understand gender norms, power relations, structural inequalities: those are issues that I don't think everyone can quickly become well versed in. Just in the way you would look for an agronomist if looking at soil, you need some expertise for studying gender – the right kind of expertise for what you want to know. I have made a point of making that analogy when speaking to non-gender scientists about gender research, of turning the tables and acknowledging that I lack expertise in their area and that I recognise the value of their expertise, and letting them come to similar conclusions on the other side. I don't feel qualified to do the genetic work or the soil work, but the other way around doesn't make much sense either.

From a practical point of view, does this mean that it is more effective if gender specialists train other social scientists, rather than having a geneticist or other technical scientist do gender analytical work? Is this more about juxtaposing and valuing different peoples' knowledge and expertise?

Marlène: I feel very strongly about that. We don't need to bring the breeders to the same point or place as the people working in the field [with communities] or with social data. That is not the breeders' job but it is important for them to see how it all fits together and to have that in mind when thinking about what they are doing, who for, and why it matters. It does our profession good too. I worked many years to get here – there was a whole process of learning and preparing for this work. We can't expect others to get there without a significant personal commitment or journey. Capacities for doing gender analyses need to start before we even look at gender, in how we understand knowledge production, in the basis of the social sciences…it is about how we know the world. That's also what I mean about being able to converse and have some fluency in thinking across the biophysical science–social science divide – if people don't recognise that knowledge is situated, if they don't question the notions of 'objectivity' and 'subjectivity', then little about the social world makes sense. Ideas about gender and social relations will be floating, not grounded. The conversation needs to start much earlier; the foundation [for thinking in this way] is much deeper.

Let's get the right people [for doing gender research], and the rest to see the big picture, be gender-responsive, and see the value and necessity in collaborating.

What would enable Marlène's team to get to the point of fluency whereby they can recognise the situatedness of their perspectives, and begin to move forward through dialogue across disciplines? For this Marlène turns to experiential learning in a field context.

Marlène: For me, everything is consolidated in the field, when you start to see things around you. If you have new ideas as a result of a dialogue or reading, take them to the field, and start to observe in a different way. After that, when people come back from the field, they say: I hadn't noticed that before – how long that woman had to walk to collect water or fuelwood…It is about bringing an awareness of gender issues into the context where you work and how this maps out there, in a field setting.

Bringing together and connecting peoples' own experiences and academic understandings, whilst engaging with a field context is an approach that is closely associated with field-based critical pedagogy being practiced in universities either through global experiential education programs (as described by Oberhauser, 2019) or in what Cravey and Petit (2012) refer to as learning through the body and through place in the field, which, advocates suggest, provide opportunities for transformative understanding.

Within the context of BI's project teams, learning through a field context not only enables people to validate their existing knowledge and expertise, but, by bringing issues into view through places and communities, it is also an opportunity to acknowledge the circumstances of privilege that Ferguson suggests makes Freire's (2000) 'pedagogy of the oppressed' difficult when using it amongst relatively privileged researchers and practitioners (Ferguson, 2019). Where gender training is based on people reflecting on individualised personal experience (i.e. through emotional bodies), this can further entrench and normalise masculine, white or colonial privilege unless these are directly confronted (Cornwall, 2016). A fieldwork context provides a more relational context that includes other axes of experiential knowledge through which people can reflect.

Marlène: I remember when I was in university studying for months in the field, in rural settings in Kenya, with some colleagues. Every night we talked a lot, we couldn't sleep. Our heads were filled with so many ideas and questions. The field is an ideal situation. If your project team is in the field, get them talking, discussing over meals, observing, seeing things in new ways. So much is informal, that is when things start to come to life, when you see things. When you have an emotion associated with a piece of information, you learn it.

Whereas her team is quite small, which is a manageable sphere of influence, an issue that Marlène is acutely aware of is that donors want to see an upscaling of change, and as she puts it, 'we still have to look for ways to influence small but radiate big outwards'. It is to those thorny questions of the logics of knowledge generation at scale and with donor agendas in mind that the final section of this chapter now turns.

Working with the logics of technocratic research and impact evaluation

When Marlène first arrived at Bioversity International, much of her effort was directed towards making existing projects, which were largely technical, more gender-responsive. As time has gone on, this has begun to shift, as she has been more involved in project development from the outset. However, she points to the logics of project development and evaluation of impact that work against the more transformative forms of learning to 'do gender' in ways that are meaningful for research teams and the communities in which they work. Three issues emerged in our discussion that are pertinent to the logics of research practice at BI. These resonate with other similar organisations working internationally and within the context of a development industry that is managed through technocratic instruments such as log frames and quantitative forms of impact evaluation, as other chapters in this book suggest. These work as structural disincentives for leveraging the kinds of transformative learning in research practice that we reflected on in the preceding section.

First, Marlène notes that as a gender specialist, the requirement to work internationally and with teams in many different countries in BI makes it difficult to deepen engagement in specific projects with particular geographies, or even to be able to build synergies across different research initiatives that might broaden and deepen understanding in and of a given place. While she stresses that being able to work in different parts of the world is a privilege and that it offers scope for building a more comparative perspective, she compares this universalising mode of working with the deep knowledge and understanding that she was able to develop through her doctoral research.

Marlène: My reality is very different [from her days doing her doctoral research]. I'm in many different countries, I'm in different teams working on different thematics. If you work in four regions, your ability to master a literature that is geographically located is impossible, you can't do it justice. You have to collaborate with other people in your field who are very knowledgeable about a specific place. In that situation, we can each bring different contributions. It is not necessarily interdisciplinarity, but it is a different kind of productive dialogue with people who are coming from different geographical experiences.

Such an approach would be a small step towards challenging the building of professional gender expertise on a universalised gender knowledge, assumed to be readily transferable and applicable across contexts (Kunz et al., 2019; Narayanaswamy, 2016).

Secondly, Marlène notes the ways in which technical priorities can dominate problem definition and research design, even as there is an awareness of how this might be at odds with responsive applied research.

Marlène: In the teams I work with, I think there's a respect and appreciation, at least in the team set-up, for having a better balance of social science and biophysical science inputs, and directing the research questions from these different perspectives. A CGIAR gender colleague suggested that it should be the social scientists driving the proposals...because the problems to be defined are by nature social science problems. The kinds of challenges we are dealing with are about poverty alleviation, food security and improved management of natural resources. Often what happens is not only a technocratic retrofitting of gender analysis into a technical field, but also people coming in to a proposal with what they know, what they do, building a work package and then trying to build a narrative around how it will help solve a problem – so, trying to make their 'solution' relevant to a problem, rather than starting from the problem and figuring out whether or how their work is relevant to it.

Seemingly, the transformative learning 'moments' that Marlène described in the previous section are difficult to leverage within prevailing technical project management requirements, whether at BI or in other environment and development organisations (Li, 2007; Mosse, 2011), and in particular where epistemic authority is more readily tied to science and quantitative social science disciplines (Kunz et al., 2019). Yet knowledge hierarchies amongst individual team members from biophysical and social science disciplinary backgrounds are not the only obstacles to address.

Marlène: The framework within which we develop projects, for example, the requirements of donors, the use of log frames, the requirement for measurable indicators: it is a particular logic that suits very well certain types of knowledge generation fields and disciplines over others. So it is not only about who gets to define a research problem, but the frameworks within which we are operating. It is also the time frame within which some of those impacts have to be delivered that does not favour certain kinds of research.

The dangers gender experts face in establishing their authority by working through governing technologies that 'reflect the episteme of the biophysical and natural sciences' (Mukhopadhyay & Prügl, 2019: 10) is that these close down opportunities for less linear and predictive research designs and privileges certain methodologies (e.g. surveys and quantitative research) over ethnographic or other qualitative approaches deemed overly 'subjective'. The latter require sustained engagement with people in places, that may generate the most meaningful gender analysis, and through this, opportunities for transformative change.

Thirdly, the time frames under which research and its impact are 'measured' within project logics work against a more transformative agenda in gender-responsive or gender-focused research. Here, despite the epistemic power accorded to the more biophysical disciplines with which Marlène works, she has found surprising common ground and a convergence with technical colleagues. They have shared their frustrations with project cycles and the timelines for delivering research outcomes and impact within the logics of donor project frameworks. In breeding programs, for example, projects may have a 20 year run when starting to breed for something and before there is measurable impact. Discussions around this have contributed to a better understanding between social science and technical colleagues.

Marlène: I heard one of my colleagues say that when he works on trees it can take him 15 years to get some data, as opposed to social science where you 'just go out and do a survey'. That was very interesting because that's how I feel too [about short time frames for making a difference]. Seeing any kind of change takes a long time in society and in social relations, so being expected to report in three years on how you changed gender relations is not a reasonable request.

Other measures of success that Marlène has to consider include wider evaluations of the gender mainstreaming program itself, and in particular, how quantitative data on this may be used to evaluate impact. This can include the amount of funding set aside for social science and gender research, compared to funding allocated for other types of research within CGIAR Research Programs. In most CRPs, very substantial amounts of funds are directed towards breeding (which is expensive to execute) whereas a gender research project might require significantly less funding but in reality, be very impactful.

Marlène: There's so much money in breeding compared to everything else. We previously had to report year-on-year on gender budgeting to show the gender responsiveness of our research portfolios. If only ten percent of the project budget is on gender research, is it gender responsive? How meaningful is that measure? If the gender research has steered the whole program, it is not about having the same money but about influence. For example, it can be about being able to influence how the project is going to be run and what you are breeding for, how a new breed fits with the bigger picture in terms of influencing gender relations and so on.

The suggestion that 'impact' is made intelligible when it is given a monetary value is typical of the market logic that underpins project evaluation within the dominant development vocabulary (Davids et al., 2014). Given these issues, we were interested in discussing whether there were differences in perspective over what kinds of change BI was working towards, and whether such differences also created a structural disincentive for meaningful and transformatory gender work. In particular, how is the link between research impact and increases in productivity being understood, given the criticism of this from feminist political ecology (see Chapter 6)?

Marlène: In this organisation the focus is biodiversity management and conservation, so if there is emphasis on productivity, it is through biodiversity. For example, if you are adding biodiversity to a cropping system to manage pests, you may need to show that you are also increasing or at least not creating any losses to productivity. From a gender perspective, you would question whether productivity should necessarily be a goal to begin with. As critical as I am, when in the field I hear a lot of women and men farmers saying it is extremely important to them. Obviously there are many different aspects to that story: questions around labour, markets, and when you think that more than 30 percent of food that is produced is lost along the value chain, you wonder if productivity is really where so many efforts should go – especially if it takes 10, 15, 20 years to breed a crop with a higher productivity. At least the CGIAR is broadening its vision beyond productivity now by talking about 'sustainable food systems'.

What about other forms of change, and in particular attitudes towards gender transformation within BI and beyond? As Marlène notes, CGIAR Research Programs and centres need to demonstrate commitments to gender mainstreaming through their gender strategies, research, and inclusion of gender specialists among their scientists. But what about change in terms of gender equality within the communities in which BI works? Is there some hesitancy there?

Marlène: This is something I come across all the time. Some colleagues say: 'we don't want to interfere with people's lives', that changing their productivity is not the same as changing social values. There are different ideas about what kind of interventions are acceptable. Gender research strikes at the core of societal values, and according to some colleagues it is hegemonic – we are imposing our norms on how societies should be. It is good that these issues surface, but they [those within the CGIAR and BI expressing this view] don't think of their own work as being intrusive. They don't see that a commitment to raising productivity or encouraging market integration is also pushing a certain norm about how a society should be.

Put bluntly, scientific research for development and a universalising approach to gender equity are both part of the same hegemonic development discourse that narrows social development concerns to particular issues (e.g. productivity, market integration). In that respect, singling out gender research for the charge of intrusion and changing people's values is misplaced (Narayanaswamy, 2016). However, there is opportunity in the slow and contextual approach to knowledge and transformative learning that we have discussed in the preceding section for deeper reflection, which might include consideration of a universalising coloniality in development *and* gender research. Through such reflection, researchers and practitioners might consider ways to push against the logics of environment and development practice where these elide or even damage the diverse interests and concerns of women and men in different communities.

Conclusion

Within a research-for-development organisation such as Bioversity International, the role of the gender specialist is expansive and complex, embracing critical pedagogies to build effective multi-disciplinary teams whilst navigating the practical, political and technical challenges set by the logics of donor-driven project design, execution and evaluation. As Marlène notes, gender professionals in this field are tasked not only with situating complex social systems within the environmental and poverty contexts that mutually constitute them, but also with 'transforming' these socio-environmental systems. Transformation must occur within a short temporal scale that is not commensurate with the time required to facilitate lasting social change. The social change must be measurable, often quantitatively, using indicators that do not adequately capture processes such as empowerment, and that may leave little room for understanding the significance of change from local women's and men's own perspectives.

The task at hand for BI (and the CGIAR more broadly) is also to produce models or approaches that can be replicated across geographies, sometimes globally, despite the context-specificity of social relations and cultures. This must happen in interdisciplinary teams where members from different disciplines, cultures and genders negotiate their own social relations, and where some team members resist the idea of gender equality as a research outcome, based on a reluctance to intervene in the 'private' lives of project 'beneficiaries', not recognising that the technocratic fixes they propose already do. Marlène's insights have revealed some of the paradoxes of gender research in an applied context, where the expectations of donors that fund the research for development system run the risk of hindering the possibilities of conducting quality, qualitative research on gender relations from which meaningful social change can be built.

Where Marlène has been able to point to success are those relational instances with colleagues and research teams in which a process of discovery and 'learning with' enables long-lasting changes in approaches to research. The visceral and embodied forms of learning that are possible in the field where people are able to see the world differently, and gently adjust their sensitivity to questions of gender equity chime with approaches to critical field-based pedagogy, and offer opportunities for dismantling epistemological divisions within multi-disciplinary teams. For Marlène, leveraging these transformative moments will involve a greater commitment to 'slow research' that includes opportunities for deepening contextual engagement, embracing different forms of expert knowledge (across disciplines, across geographies) and critical reflection with others. This means going beyond developing a common scientific language and rendering gender technical in the process, but instead taking seriously other ways of working and forms of knowledge in 'agriculture for development' (A4D) that is meaningful for people and their communities.

References

Arora-Jonsson, S. (2014). Forty years of gender research and environmental policy: Where do we stand? *Women's Studies International Forum*, 47, 295–308.

Arora-Jonsson, S. & Sijapati Basnett, B. (2018). Disciplining gender in environmental organizations: The texts and practices of gender mainstreaming. *Gender, Work & Organization*, *25*(3), 309–325.

Bustelo, M., Ferguson, L., & Forest M. (Eds.) (2016). *The Politics of Feminist Knowledge Transfer: Gender Training and Gender Expertise*. London: Palgrave.

Chambers, R. (1993). *Challenging the Professions: Frontiers for Rural Development*. London: Intermediate Technology Publications.

Chambers, R., Pacey, A., & Thrupp, L. (Eds.) (1989). *Farmer First: Farmer Innovation and Agricultural Research*. London: Intermediate Technology Publications.

CIFOR (2013) *Gender in the CGIAR Research Program on Forests, Trees and Agroforestry: A Strategy for Research and Action*. Bogor: Center for International Forestry Research (CIFOR). https://www.cifor.org/knowledge/publication/4056/ [accessed 5 June 2020].

Conway, G. (1985) Agroecosystem analysis. *Agricultural Administration*, 20: 31–55.

Cravey, A. J., & Petit, M. (2012). A critical pedagogy of place: Learning through the body. *Feminist Formations*, 100–119.

Davids, T., van Driel, F., & Parren, F. (2014). Feminist change revisited: Gender mainstreaming as slow revolution. *Journal of International Development*, *26*(3), 396–408. https://doi.org/10.1002/jid.2945

Davids, T., & van Eerdewijk, A. (2016). The smothering of feminist knowledge: Gender mainstreaming articulated through neoliberal governmentalities. In *The Politics of Feminist Knowledge Transfer Gender Training and Gender Expertise*. (pp. 80–96). London: Palgrave Macmillan.

Ferguson, L. (2019). Exploring privilege through feminist gender training. *European Journal of Politics and Gender*, *2*(1), 113–130.

Freire, P. (2000) *Pedagogy of the Oppressed*. London: Bloomsbury Publishing.

Friis-Hansen, E., & Sthapit, B. (Eds.) (2000). *Participatory Approaches to the Conservation and Use of Plant Genetic Resources*. Rome: International Plant Genetic Resources Institute [now Bioversity International]. https://www.bioversityinternational.org/fileadmin/_migrated/uploads/tx_news/Participatory_approaches_to_the_conservation_and_use_of_plant_genetic_resources_603.pdf [accessed 27 May 2020]

FTA (2020) *Gender Equality and Social Inclusion: A Revised Agenda and Action Plan for the CGIAR Research Program on Forests, Trees and Agroforestry 2020–2021*. Bogor: CGIAR Research Program on Forests, Trees and Agroforestry (FTA). https://gender.cgiar.org/ftas-new-gender-equality-and-social-inclusion-agenda-and-action-plan-2020-2021/ [accessed 27 May 2020]

Harcourt, W. (2019). Feminist Political Ecology Practices of Worlding: Art, Commoning and the Politics of Hope in the Class Room. *International Journal of the Commons*, *13*(1).

hooks, b. (1994). *Teaching to Transgress: Education as the Practice of Freedom*, London: Routledge.

Kolb, D. A. (1984). *Experiential Learning: Experience as the Source of Learning and Development*. Englewood Cliffs, NJ: Prentice-Hall.

Kunz, R., Prügl, E., & Thompson, H. (2019). Gender expertise in global governance: Contesting the boundaries of a field. *European Journal of Politics and Gender*, *2*(1): 23–40.

Li, T. M. (2007). *The Will to Improve: Governmentality, Development, and the Practice of Politics*. Duke University Press.

Marx Feree, M., & Verloo, M. (2016). Foreword. In Bustelo, M., L. Ferguson & M. Forest (Eds.) *The Politics of Feminist Knowledge Transfer: Gender Training and Gender Expertise* (pp. x–xiiv). London: Palgrave.

Mosse, D. (2011). Introduction. The anthropology of expertise and professionals in international development. In D. Mosse (Ed.), *Adventures in Aidland: The Anthropology of Professionals in International Development* (pp. 1–31). Berghahn Books.

Mukhopadhyay, M., & Wong, F. (2007). Introduction: Revisiting gender training. The making and remaking of gender knowledge. In M. Mukhopadhyay & F. Wong (Eds.) *Revisiting Gender Training: The Making and Remaking of Gender Knowledge – A Global Sourcebook* (pp. 11–26). Amsterdam: KIT Royal Tropical Institute.

Mukhopadhyay, M., & Prügl, E. (2019). Performative technologies: Agricultural research for development and gender. *International Feminist Journal of Politics*, *21*(5), 702–723.

Narayanaswamy, L. (2016). Whose feminism counts? Gender(ed) knowledge and professionalisation in development. *Third World Quarterly*, *37*(12), 2156–2175.

Oberhauser, A. M. (2019). Transformation from within: Practicing global education through critical feminist pedagogy. *ACME: An International Journal for Critical Geographies*, *18*(3), 751.

Prügl, E. (2016). How to wield feminist power. In M. Bustelo (Ed.), *The Politics of Feminist Knowledge Transfer* (pp. 25–42). London: Palgrave Macmillan.

Said, E. (1978) *Orientalism*. New York: Pantheon Books.

van der Burg, M. (2019) 'Change in the Making': 1970s and 1980s building stones to gender integration in CGIAR agricultural research. In C.E. Sachs (Ed.), *Gender, Agriculture and Agrarian Transformations* (pp.35–56). London: Routledge.

Wenger, E. (1999). *Communities of Practice. Learning, Meaning and Community*. Cambridge: Cambridge University Press.

8 Please genderise my log frame

Interactions with technical specialists for gender mainstreaming in environment projects

By Bernadette P. Resurrección and Rebecca Elmhirst
In conversation with Annette Wallgren and Victor Tsang

International feminist activism and research has had significant success in pushing gender issues onto the international agenda and into global governance institutions and processes, including in those deemed to be science-, technical- or environment-related institutions. The goal of gender equality is now widely accepted and codified in international normative instruments, such as those propelled and promoted by the United Nations, through the instrument of gender mainstreaming. Additionally, the donor and development community has championed gender mainstreaming in far-reaching ways, such that even environmental organisations are now tasked to implement it.

Since the UN Conference on Environment and Development (UNCED) in Rio Janeiro in 1992, international environment agreements – some more than others – have contained legally binding clauses and specific guidelines promoting gender-responsive sustainable development in biodiversity, desertification, and natural resource management. Following UNCED, the Fourth UN World Conference of Women held in Beijing in 1995, governments worldwide endorsed the landmark Beijing Platform for Action, where Section K is entirely dedicated to women and environment linkages that will guide future gender mainstreaming policy and action. Governments and civil society actors participating in Beijing endorsed the strategy of mainstreaming a gender perspective into all public policies in order to counter the gender bias in society and policies and to produce gender-equal policies.

The linkages made between women, gender and environment in these agreements were largely influenced and framed by proponents of Women, Environment and Development (WED) and a Global South variant of ecofeminist thinking. These gender and environmental analytical frameworks fundamentally recognise and track the dichotomous differences of environmental impacts and responses by women and men, arguing that women are hardest hit by environmental stresses and therefore are more favourable environmental agents (Resurrección, 2013, 2017; Elmhirst & Resurrección, 2012). This, henceforth, became the persistent gender mainstreaming rationality that travelled across environmental programs and policies until the present time. These frameworks landed on favourable scientific ground owing much to their propensity to focus on measurable impacts from environmental degradation, with less attention to the political economic and discursive origins of where these impacts might be anchored and possibly challenged (MacGregor, 2010;Tschakert, 2012)

In similar vein, much of the growing literature on gender mainstreaming has highlighted and warned of the ways with which it has watered down, technocratised, effectively blunting the political edge of transforming gender and power relations (Prügl, 2015; Parpart, 2014; Davids et al., 2014; van Eerdewijk & Davids, 2014; Tiessen, 2007; Mukhopadhyay, 2004; Miller & Razavi, 1998; Baden & Goetz, 1997). How then do we approach existing gender mainstreaming efforts particularly in environmental organisations by gender professionals? In this chapter, we venture a different track by reflecting on the relationships – often cautious and uneasy – that are typically forged in large part due to the efforts of gender professionals who for better or worse occupy the divides between technical science and social change imperatives. We then explore and reflect on how 'feminist strategizing develops its own rationality: (often) to govern through subtle and indirect means' (Çağlar et al., 2013: 5; parenthesis included) in an environment and development organisation.

The reflections of gender experts Annette Wallgren and Victor Tsang of the UN Environment Programme (UNEP) tell us of the myriad ways of navigating a conventionally technical and biophysical knowledge and policy terrain as representatives of gender mainstreaming. By doing so, they unravel how gender and environment regimes are often co-constituted by the social practices of the gender experts and technical specialists who co-inhabit, discipline and define the logics of this terrain. From their accounts, but without essentialising the ways in which their own positionalities play out, we are also able to explore the different ways each of them is navigating the terrain.

Becoming gender experts

Twenty years ago, Miller and Razavi (1998), in their book, *Missionaries and Mandarins: Feminist Engagement with Development Institutions*, told us that feminist incursions have been greater in institutions or parts of them that are generally amenable to gender and social justice concerns. The International Labour Organization, for instance, is one of them. Institutions where economic efficiency and growth are the chief concerns such as in the World Bank, the landscape was free from feminist incursions. However, not even a cursory mention about feminist engagements with scientific and technical organisations was present in Miller and Razavi's book, as these were probably deemed too remote from the radar of feminist activity. Things have significantly changed since then.

The first full-time gender experts in UNEP were appointed in 2007. Since then junior and mid-level positions have been created for gender experts across its offices worldwide. Victor Tsang is stationed in the UNEP headquarters in Nairobi, while Annette Wallgren is based in the Asia Pacific regional office in Bangkok.

Victor: UN Environment Programme works on a variety of issues to protect our shared planet, from the promotion of renewable energies, minimising the harmful effects of chemicals to protecting species in forests and oceans. We are unique as an UN organisation as our headquarters is located in Kenya, a lower-middle-income country. In practice, UNEP carries out assessments, convenes experts in workshops, organises trainings for governments, sets standards, builds alliances across sectors and sometimes engages community members in demonstration practices. It works primarily at the normative level. Many of our colleagues are scientists, bureaucrats, engineers and economists. There are also political scientists, social scientists, journalists and people in support functions such as finance and administration. There are more men than women in professional positions, especially at higher grades. Colleagues in general maintain a positively friendly and healthy relationship, keeping professional distance. There was a time when we could not imagine full-time gender professionals in UNEP, and it seems the enthusiasm for gender mainstreaming is growing.

Annette: We have women in our top management positions and I think that is generally well accepted. It is a friendly and open environment, it is easy to consult with colleagues. Gender has become more recognised as a core business.

Babette: Why do you think gender issues have found their way into UNEP? Was it because of the momentum caused by Agenda 21, which was the outcome of the agreements made in Rio de Janeiro in 1992?

Victor: At that time of the Rio Conference in 1992, the concept of sustainable development very much focused on environment, which is quite different from how it is generally understood today. The economic, social, and environmental dimensions of sustainable development have become more integrated, as there's in fact a push for people to begin to see the link among these things. I guess, in the past, people tended to see that gender is more important for development but maybe less for the environment. So, I guess we need to present the links between gender and environment in a different way for an environment-focused audience. For example, we can say that changing environments have different impacts on women and men, so we need to ensure that both are involved in decision-making regarding the management of resources.

Annette: Just by talking to colleagues and hearing them say things like, 'Everything is about gender now [laughter]', makes me think that maybe it's a sign that gender is now gaining more importance. Because they have been in this business and worked in UNEP for 20 years.

Becky: Given the heightened recognition of gender issues in environmental policy contexts, what do you understand by the term 'gender expert'?

Victor: It means someone advancing the cause of gender equality as fully or partially her/his profession, with formal terms of reference. The concept should be differentiated from 'women professionals', or people who promote gender equality on a voluntary/informal basis, or 'gender champions', or people working in a women's organisation.

Annette: I also see it as a very technical expertise, which it is not always valued as such, as Victor is saying – sometimes more as an advocacy role.

Babette: What was your disciplinary background, and did this help you attain the status of a gender expert? Sometimes we are hired not by our educational background but by our work background.

Victor: I had a master's degree in Education, Gender and International Development from University College London, and had recently completed two study modules with the School of Oriental and African Studies on sustainable development and climate change. My degree in gender has been instrumental, especially in the earlier stage of my career when I was with the World Food Programme (WFP). The confidence and learning capacity acquired from the degree helped me pilot outcome-level gender indicators for WFP. During 2013–14, I was able to develop, pilot-test and get approved WFP's first ever corporate gender outcome-level indicator. I grew up in Hong Kong, a traditional society where there were set roles for girls and boys. I worked in an NGO in the rural villages and communities in northwest China – villages, communities. When conducting a survey in a rural area, I was knocking on someone's house and asked whether there was anybody home, and a woman came out and said 'no.' I was first curious and later became puzzled, why a woman would think of herself as no one, and that was how I started my interest in gender.

Annette: I did my studies in International Crisis and Conflict Management, and later a Master's in International Politics, specialisation Peace and Conflict and all my independent research and assignments focused specifically on gender roles. For example, I did research in Ethiopia on gender roles in times of crisis, where I learnt how the Eritrean refugee men in Ethiopia suffered a loss of masculinity after losing their ability to provide support to their families. Ever since, the notion of men's as well as women's perspectives have been central in my work. I also did research in Kosovo looking at women returning after the war to rebuild – as political actors – the country after the conflict. I also did gender studies in Canada. My consciousness for gender comes from the activist side of me in my younger years, realising early on in life that there are inequalities – expectations foisted on boys and girls – but also realising that these structures can be changed.

Babette: You both seem very qualified for the profession, with great background experiences. How do your present colleagues respond or react to your professional role within the organisation?

Annette: (Laughter) It's almost expected from me because I'm from Sweden. However, I often come across comments like 'There are no linkages to gender in this project', often due to lack of capacity and understanding. Then it is important to support colleagues in unpacking the issues and identify the linkages between gender and the particular environmental problem that the project is addressing. This is the first time our regional office has a full-time gender professional so within these initial years I had to find my role, establish a portfolio and make a case for the need for gender inclusive programs.

The fact that this position is more junior also has an impact, and if I'm cynical, I'd say perhaps I'm not as influential. However, starting from scratch, we have now established a five-year project on gender equality and climate change, gender responsive biodiversity planning projects in two countries and work areas of climate change law, chemicals and waste, forestry, fisheries and resource efficiency that include gender in a more systematic and responsive manner. As a result of this portfolio, my role as a gender professional has become clearer within the office and my voice and authority have emerged.

Victor: I usually present myself as working on the 'people' side of the environment. My male colleagues talk to me how they really feel about including gender in their work. The chance is also bigger for men to open up to me. It happens that they will confess to me their 'politically incorrect' thoughts, such as their disagreement with the UN's gender parity policies. I position myself as a social scientist, a gender specialist, and a friendly colleague and sometimes as a friend, so that I have the legitimacy to provide technical and institutional gender support in a positive atmosphere.

Apart from the earlier international environment and development agreements that have carried gender-inclusive languages, over the course of recent decades, gender equality has evolved into an institutional imperative even in technical organisations (see Introduction). Gender mainstreaming has become the central political strategy of environment and development organisations to address gender inequality: thus, in a sense, gender mainstreaming has now become 'top down' and has acquired its own rationality from discourses and social processes within which these organisations are embedded.

Gender experts, more than ever before, are populating the ranks of professional bureaucrats in multilateral international environment organisations and have skilfully stepped in pace with the architecture of growing information systems, technologies and tools for assessments, data capturing, visualisation, modelling and measuring. The UN is particularly tasked with creating norms for gender equality among national governments for their policy making, and along these lines, UNEP purports to mainstream gender in its programs, networks and engagement operations.

Gender professionals within the UN bureaucratic machine often struggle with patriarchal norms and practices (Sandler and Rao, 2012; Caglar, et al., 2013; Miller & Razavi, 1998). Joanne Sandler and Aruna Rao (2012), for example recall the sexist and misogynistic remarks that were used to belittle the efforts of men involved in gender mainstreaming activities in UN organisations. On the other hand, Miller and Razavi (1998) point out that gender experts are the 'feminist policy entrepreneurs' who draw attention to the innovative nature of their work, identify new issues, skilfully mobilise facts which can justify action and cultivate internal allies and champions. Fellow bureaucrats also become allies with gender experts based on multiple deflections of identity – gender, nationality and race – as Annette and Victor hint at, cultivating webs of relations that can push the gender agenda forward in potentially productive ways. The next sections will show how gender mainstreaming in UN organisations can create conducive spaces for dialogues albeit tainted by instrumentalist manoeuvrings and 'acting pragmatically to secure small instrumental changes' (Eyben, 2010). Some quarters remain intransigent to change, while others signpost a 'slow revolution' (Davids et al., 2014).

The challenges – and golden moments – of gender mainstreaming

Babette: I'm sure you've devised strategies on gender mainstreaming. They can be very unique.

Victor: As a mechanism for accountability for gender mainstreaming, I introduced the Gender Marker system to UNEP in 2015, and I reviewed all new projects according to the Marker's four criteria. Through the compulsory project review mechanism, I am able to provide comments and suggestions for improvement. In rare cases when the projects are gender-blind, I confront the project managers, being firm but still trying to help in a friendly way. For the gender marker, we use a four-point scale. A score of 0 means gender was not applied at all. A score of 1 and 2a indicates that there was some application. And then sometimes they score 2b, which means that the project is a gender-targeted initiative. So usually, we have this four-point scale, but the criteria will probably not be the same for a different organisation, I think.

Annette: My starting point is always the gender analysis, which will guide the project. This is where colleagues and I start unpacking the environmental problem to identify the gender related areas, situation for men and women, who is impacted, who should be part of the solution, where (data, existing policies, publications, research) and with whom can we find the information we need (who to consult: women's groups and NGOs, men and women in the communities, women's ministries, etc.) This gender analysis is often missing, which means that it becomes difficult to include gender in implementation and measuring impacts. Some projects I have seen move directly to sprinkle some gender activities in project implementation without a sufficient gender analysis, which is a big issue. The risk is that we implement activities based on assumptions and without understanding the specific context and the gender roles; secondly, it may not make sense to those in the project team who are responsible to implement the activities – leading to further resistance of gender mainstreaming.

Babette: Is gender work or mainstreaming almost like a skilled service? Sometimes it feels like we're handmaidens of mainstreaming (laughter). You know, as service providers.

Annette: It's been an interesting process since I started. Since it was a new position, no one had been a gender officer before in the office. It was a bit daunting to try it out. What's going to be my role? I needed to sort of justify (laughter) or build up the profession itself. So, in the beginning, it was a lot of this advisory work in all areas.This advisory role has not stopped but it might have decreased a little bit because colleagues have actually started to pick up on gender, so I don't need to always be involved. And there is a larger number of gender champions and they take initiaives themselves. I think I'm going to steal a quote from one of our partners that was said over dinner this week. But Victor might help me to say exactly what she said. She said something like, 'Can you genderise my log frame?'

Victor: Yes, exactly: 'genderise my logframe'. And sometimes, you receive an email where someone is asking, 'Can you please add some activities here and sprinkle some gender gold dust over this project document?'

Annette: But what is interesting then is that I don't mind having people asking me and seeking advice. I think that it's also what I really enjoy. But through this other work of an established gender project portfolio, I think I've gained a bit of more…

Victor (suddenly stepping in): Respect.

Annette: Yes. So, I don't feel that it's the same as when I started. They don't see my job position as a supporting role only anymore. It's more a professional, technical expertise.

Becky: Since we're now talking about change, in your experience what were those breakthrough moments, and what led to them? Even if you say that it's on the mainstreaming front, what made them golden, and why do they indicate some kind of breakthrough?

Annette: I guess the breakthrough happens when colleagues themselves conclude how and why it makes sense. I don't have a gender marker, or I don't have that much of a stick to use (unlike Victor), so what I sometimes invoke is the donors' preference to integrate gender issues, or the time dedicated for gender concerns as allocated by funds, or whatever requires gender to be included. But I think the best approach in my work is to use the softer (laughter) approach which means that, okay, we sit down, and we talk, and this colleague says, 'oh it doesn't really make sense here and I think it's a bit too much and it's not relevant, my work doesn't really relate to gender.' So, then we just start to discuss, and I say, let's not make this so complicated, it shouldn't be so hard and let's make it work for you in this area and we don't need to go over the whole theory. Because also sometimes they say, I've done this training and I don't understand, and it doesn't make sense what they're talking about and how their ideas relate with my work. And then that's when I start asking questions like, 'so how does your project relate with people? Does it relate with people in any way? And the golden moment is when I don't have to preach, I don't have to teach, it comes after a while through reflection from their side. And since they are the experts on their work, I am not, when it makes sense to them and why it should be included, in what way it can be included that's not too complicated, that's really nice (laughter). That's really the breakthrough moment. I'm not a lecturer or I don't prescribe to them how they should do it, it's really about getting them to understand for themselves why gender is important. And this is a strategy that works for me, I think.

Victor: Also, we shouldn't be judgmental. When people make some jokes, even if they're wrong, we don't react immediately. If people feel that they're being judged, they will switch off and then we lose the opportunity to continue the relationship with them. The approach should be friendly and positive, keeping things simple, and enabling people to feel ownership over their work.

Facing resistance and fear

Becky: Do you often face resistance since you work in a highly technical environment?

Victor: From time to time, there is resistance, since no one wants to do additional work. Gender mainstreaming is still considered additional. Despite their good intentions, they often really don't know what to do, then that's when they find it demanding. Sometimes they react and ask why we don't focus instead on disability, as they are not convinced that gender is the most important consideration. They see it as something political.

Annette: The task of gender mainstreaming also comes from management direction, from a higher level, through our different screening systems, so it's not just coming from me, but it's broader. This is perhaps why people don't complain regularly, although there is some resistance as Victor says. This results in people wanting me to do the job, and in the beginning I did that so that I took on a lot of extra work such as adding gender dimensions to documents and project designs. I had to delegate it back but provide support.

Victor: A colleague who's very knowledgeable and has long worked on biodiversity ecosystems said a few times that she really enjoys working with you Annette, as you make it understandable for her. She also explained that, 'In my previous work, the gender people I met have been sometimes making things very complicated.'

Annette: The same person also seemed to hint that some gender professionals she's known have been quite confrontational and pushy and she thinks that Victor and I are quite nice and want to make it easy for people. And that we don't want to complicate it too much and overwhelm them. But there's a balance to not reduce the importance of gender. This is something I sometimes struggle with. I want it to be convenient for my colleagues, but I also want to make them understand that this it actually does need some work. It does need some effort. It doesn't come easy, and it's not a quick fix.

Babette: Do you do a lot of persuading?

Annette: When I meet them one on one they share their lack of knowledge of and how to integrate gender issues in their work. People don't like being exposed and are afraid of it. For instance, when we discuss climate change adaptation, we try to unpack the roles of women and men as they adapt. I ask them questions – what do men and women do? Are their livelihoods impacted due to climate change? What similar and different constraints to they face? I try to make gender concrete and understandable and create a relationship of trust.

Resistance to gender mainstreaming and gender training has been studied closely by Lombardo and Mergaert (2013). They cite several reasons for people's resistance. For instance, they point out that gender mainstreaming efforts challenge people's personal identity and personal beliefs, sometimes making them feel exposed to criticism or suggesting transforming practices that they had thought were conventional. Questioning the personal sphere can also trigger fear and self-protection that can cause resistance in people. Another reason for resistance may be that people consider gender equality and feminist goals to be ideological and emotional, rather than rational, scientific or legal. They also resist when they feel that trainers are trying to change their attitudes or manipulate them in some way. Others may also deny that there is need for change in the existing order of things, referring to the 'natural or innate' order of things otherwise, advocating that gender inequality is not a problem in some societies.

Annette and Victor's accounts instruct us slightly otherwise. The institutionalisation of gender mainstreaming in the UN as a liberal institution appears to be increasingly deeply entrenched, and as Victor narrates, mechanisms are in place to hold people accountable to meet gender-equal goals. In turn, it appears that people may be resisting what they perceive to be 'ideological' or 'political' additions that seem incompatible with their technical workloads. However, to openly resist or plead ignorance on gender concerns may render some of them deeply vulnerable and epistemologically not in step with a UN core value and required skills portfolio. This, however, does not suggest that patriarchal attitudes and practices are completely on the wane. Lombardo and Mergaert (2013) also observed that resistance takes many forms: such as despite strong language supporting gender equality and mainstreaming, resources are in reality scarce to make it happen, thus often times a case of everywhere, but nowhere (Tiessen, 2007).

Epistemological questions

Becky: What about integrating gender with technical environmental topics – has this been challenging as well?

Annette: I can also have a hard time making the connections between gender and environment. For example, my colleagues have an air pollution project and I found a few entry points that's for me very clear. But there's still some parts that I find can be a long shot. I do wonder, 'Does it make sense?' I have been talking to a colleague who's working on a partnership for clean air in Asia and what they're doing is having a number of member states monitor the level of pollution in the air. Which is a fantastic and important initiative for the environment and to see where the worst pollution is. So, they have a good network and are meeting regularly. And he is really keen on gender issues and would like to include that into his work. And I, of course, every time I hear this, get very excited. But to me, then, when I started to look into this project, which is not about people, but very much about data and monitoring, it seems to be a bit of a long shot. How do we include gender into air quality monitoring project?

And I think that this is why we try to draw the links between gender and environment, because if we do, then the outcome will be 'something else'. For example, I mean we can think of how environment interacts with people, or the human use of natural resources. It can become 'something else,' and not just pure monitoring of air quality. We could turn this into a gender awareness project included in there.

Victor: And also they tend to be very practical, they want something quick and easy. They are busy and people who are from a science background do not learn by reading word by word. They learn by having numbers and formulas and shapes and graphs.

We change as we work for change

Becky: Isn't there a growing need to demonstrate positive impacts from your gender and environment work?

Babette: Indeed, the funding environment nowadays emphasises that programs yield positive results on empowerment and gender equality.

Victor: We work a lot on internal capacity, standards and processes – all central to gender mainstreaming – but we hardly have any clues what impacts they lead to, and we hardly have opportunities to work in the frontline, especially in my case as I am based in the headquarters.

Annette: I have been thinking about this a lot, since some of my previous work experiences focused more directly on women's empowerment and gender equality. UNEP's goal is rarely to achieve gender equality in itself. Rather, we are making sure that we are equally involving everyone's perspective, needs, contribution, capacities. The mantra I keep sharing with my office is that gender inclusive programs creates better solutions and more sustainable environmental protection.

I think I've done a bit of a professional shift, since I started in feminist organisations where the end goal was gender equality and women's empowerment. So then moving to working in an environmental organisation, I really had to adjust. I realised that I have had to change a bit myself. Colleagues who are experts on biodiversity and climate change have taught me so much of how the process works for governments. They have made me realise that for gender work to be successful in this area, we need to understand where the governments are coming from, and their processes. So we cannot come and push gender equality too far, because then we will lose their attention, and we will lose the efforts we've put in building a relationship with them. So I feel that I have learned to work towards more realistic results.

As our work and engagements change so do our identities. 'Feminism' and 'feminist politics' in international organisations are also in flux and changing as institutional contexts change. They are fundamentally shaped and constituted by the very nature of our engagements, our evolving knowledge, the institutions that we are part of, and gender inequalities in wider society. While our ethical commitment to equality and empowerment drive us to make career choices as 'missionaries', we are also defined by the institutional and contextual boundaries within which we enact our commitment. 'Institutionalised and power-laden environmental management regimes have led to the emergence of new kinds of people, with their own emerging self-definitions, understandings of the world, ideologies and behaviours' (Robbins, 2012: 216).

This chapter ends by underscoring the contested nature of gender mainstreaming as revealed by the experiences of Annette and Victor, and by the gender mainstreaming literature in general. Gender mainstreaming processes incorporate the complexities and necessities of forging relationships and alliances. In technical organisations, we observe a recourse to 'rendering gender technical', to paraphrase a line from the work of Tania Li (2007): in order to dexterously fit the feminist agenda within bureaucracies that are immersed intractably in their own procedural inertia and cognitive mindsets, gender professionals have to skilfully simplify the otherwise complex understandings of gender and power. In the end, gender mainstreaming regimes turn out to be disciplined sites with logics derived from gender professionals' creative 'genderising': or 'bending and stretching' that are necessary to embed gender equality agendas in existing bureaucratic contexts (Verloo et al., 2009)

The mandate of multilateral technical organisations like UNEP is to inform and influence states on environmental protection. To do this, they produce knowledge products that are envisaged to potentially inform policy decision making. Annette pointed out that air quality monitoring, for example, stood on its own and linking it with gender and social issues is challenging, and perhaps even pointless. But by otherwise alluding to evolving 'something different', Annette hopes that they will produce information that is no longer free-floating balloons in a vast sky of knowledge, but that one day it will make sense to link them with actual people's gender-unequal lives.

Annette and Victor use deep engagement with their technical colleagues, which requires reflection and dialogue, tracing where and how knowledges can meet. These are some of the many travails of gender mainstreaming in an environment organisation. As Eyben and Turquet (2013) express eloquently in the blurb of their book: 'Every day, in international development organisations feminist bureaucrats make use of strategy, tactics, wisdom and skill to act for their principles. Most of their strategies are invisible and their tactics subtle. They draw on networks of friendships and relationships that create ripples of effect in enabling their organisations to be pathways of women's empowerment.'

Victor Tsang was a Program Officer of the UN Environment Programme (UNEP). He was a passenger on the Ethiopian Airlines flight to Nairobi which crashed on the morning of 10 March 2019 near Addis Ababa. There were no survivors.

References

Baden, S., & Goetz, A. M. (1997). Who needs [sex] when you can have [gender]? Conflicting discourses on gender at Beijing. *Feminist Review*, *56*(1), 3–25. https://doi.org/10.1057/fr.1997.13

Caglar, G., Prügl, E., & Zwingel, S. (2013). Introducing feminist strategies in international governance. In G. Caglar, E. Prügl, & S. Zwingel (Eds.), *Feminist Strategies in International Governance* (pp. 1–18). Routledge.

Davids, T., van Driel, F., & Parren, F. (2014). Feminist change revisited: Gender mainstreaming as slow revolution. *Journal of International Development*, *26*(3), 396–408. https://doi.org/10.1002/jid.2945

Elmhirst, R., & Resurrección, B. (2012). Gender, environment and natural resource management: New dimensions, new debates. In *Gender and Natural Resource Management: Livelihoods, Mobility and Interventions*. Routledge.

Eyben, R. (2010). Subversively accommodating: Feminist bureaucrats and gender mainstreaming. *IDS Bulletin*, 41(2). https://doi.org/10.1111/j.1759-5436.2010.00123.x

Eyben, R., & Turquet, L. (Eds.). (2013). *Feminists in Development Organizations: Change from the Margins*. Practical Action Publishing.

Li, T. M. (2007). *The Will to Improve: Governmentality, Development, and the Practice of Politics*. Duke University Press.

Lombardo, E., & Mergaert, L. (2013). Gender mainstreaming and resistance to gender training: A framework for studying implementation. *NORA: Nordic Journal of Women's Studies*, *21*(4), 296–311. https://doi.org/10.1080/08038740.2013.851115

MacGregor, S. (2010). Gender and climate change: From impacts to discourses. *Journal of the Indian Ocean Region*, *6*(2), 223–238. https://doi.org/10.1080/19480881.2010.536669

Miller, C., & Razavi, S. (1998). *Missionaries and Mandarins: Feminist Engagement with Development Institutions*. London: Intermediate Technology Publications in association with the United Nations Research Instiute for Social Development.

Mukhopadhyay, M. (2004). Mainstreaming gender or 'streaming' gender away: Feminists marooned in the development business. *IDS Bulletin*, *35*(4), 95–103. https://doi.org/10.1111/j.1759-5436.2004.tb00161.x

Parpart, J. L. (2014). Exploring the transformative potential of gender mainstreaming in international development institutions. *Journal of International Development*, *26*(3), 382–395. https://doi.org/10.1002/jid.2948

Prügl, E. (2015). Neoliberalising feminism. *New Political Economy*, *20*(4), 614–631. https://doi.org/10.1080/13563467.2014.951614

Resurrección, B. P. (2013). Persistent women and environment linkages in climate change and sustainable development agendas. *Women's Studies International Forum*, *40*, 33–43. https://doi.org/10.1016/j.wsif.2013.03.011

Resurrección, B. P. (2017). Gender and environment from 'women, environment and development' to feminist political ecology. In S. MacGregor (Ed.), *Routledge Handbook of Gender and Environment*. Routledge/Taylor & Francis Group.

Robbins, P. (2012). *Political Ecology: A Critical Introduction, 2nd Edition*. Wiley-Blackwell.

Sandler, J., & Rao, A. (2012). The elephant in the room and the dragons at the gate: Strategising for gender equality in the 21st century. *Gender & Development*, *20*(3), 547–562. https://doi.org/10.1080/13552074.2012.731741

Tiessen, R. (2007). *Everywhere/Nowhere: Gender Mainstreaming in Development Agencies*. Kumarian Press.

Tschakert, P. (2012). From impacts to embodied experiences: Tracing political ecology in climate change research. *Geografisk Tidsskrift-Danish Journal of Geography*, *112*(2), 144–158. https://doi.org/10.1080/00167223.2012.741889

van Eerdewijk, A., & Davids, T. (2014). Escaping the mythical beast: Gender mainstreaming reconceptualised. *Journal of International Development*, *26*(3), 303–316. https://doi.org/10.1002/jid.2947

Verloo, M., Meier, P., & Lombardo, E. (2009). *The Discursive Politics of Gender Equality: Stretching, Bending and Policy-Making*. Routledge.

9 The gender professional as ethnographer

Working for equitable forests

By Rebecca Elmhirst and Bernadette P. Resurrección
In conversation with Carol J. Pierce Colfer

Introduction

Ethnography is a research approach integral to the discipline of anthropology, which typically relies on the researcher's immersion in social settings through a sustained engagement with people's complex, everyday lives. Feminist ethnography pays particular attention to power and culture as played out in 'the field', and how lived experience is gendered, classed, racialised and sexualised. This approach typically disrupts other kinds of knowledges that seek to reduce human behaviour to economic rationalities, and instead emphasises meaning-making through close (participant) observation and other qualitative methods based on conversation. The relationship between development practice and anthropology (including its constituent method of ethnography) is complex and has inspired an extensive literature that has explored the way in which the field itself emerged from development's colonialist origins (i.e. its role in seeking to understand diverse peoples in order to govern them) and more recently, considers the role of ethnographic knowledge as 'evidence' in development settings (Lamphere, 2004; Crewe & Axelby, 2013; Gardner and Lewis 2015; Eversole, 2018). Ethnographic approaches are found in the tools, techniques and framings of environment and development practice, ranging from the participatory rural appraisal tools influenced by ethnographic fieldwork methods (Chambers, 1994) to a focus on the social dynamics of institutions, an example being the mobilisation of the concept of 'social capital' within the World Bank's programming in Indonesia in the 2000s (Bebbington et al., 2004; Li, 2011).

The 'equitable forest' referred to in the title of this chapter relates to the work of Carol Colfer, an anthropologist by training, who has spent nearly 25 years in forest-related research with the Center for International Forest Research (CIFOR), an international non-profit interdisciplinary research organisation headquartered in Bogor, Indonesia. As one of the Consultative Group on International Agricultural Research centers (CGIAR), CIFOR's core business is to improve forest management in the tropics in such a way that it is good for people and the environment through better policy, practice and human capacity. Carol's work with CIFOR has focused on equity and well-being of people in forest contexts, and this has involved field research, the development of appropriate participatory tools and methods, and training for stakeholder engagement. She has worked in various development sectors, ranging from forestry to agriculture to health, and whilst gender featured in much of this work, it has been over the past five years that she has been contracted specifically as a gender professional by CIFOR, bringing together her ethnographic and gender expertise. Much of Carol's work at CIFOR (and with earlier agricultural projects) has coincided with the 'ethnographic turn' (Li, 2011) that was taking place in other leading development organisations in the 1990s and early 2000s, most notably the World Bank but also USAID, where anthropologists were being hired to advance a wider agenda of community-based participatory research, including within the natural resource management sector.

In technical research environments, such as forest research institutes, ethnographers have to wear their 'boots-on-the ground' and engage with local communities, whilst at the same time collaborating with professional teams from a range of disciplinary backgrounds. The emphasis on people and institutions in community-based environment and development work means importance is bestowed on field-based social research as part of concerted efforts to discover the best means to enable people and their institutions to sustainably manage natural resources. For the gender professional working in such contexts, there is a need for engagement with diverse groups of people, from forest communities and local government, through to scientists from across a range of disciplines. Where does 'gender' fit within this process of navigating across and between different cultures (in many senses) and epistemologies in forestry research and practice? In this chapter, Carol discusses her experiences of being an ethnographer and a gender expert within a natural resources management context: both being professional identities which are drawn upon to bridge different types of knowledge and cultures to arrive at common understandings that will later guide action or enable solutions. An 'ethnographic sensibility' is key to this bridging challenge, and in weaving between the knowledges and approaches that underpin professional and academic anthropology, gender analysis and multi-stakeholder analysis in natural resource management settings.

Becoming a gender specialist and applying anthropological approaches

Carol spent her childhood in Turkey from the age of nine to 16, a formative period of her life. Her mother, like her father, was attending graduate school; her father was very supportive of women being educated and having a career, which set her family apart from most others. These years in another country, and the subsequent surprise of personal experience of gender discrimination, sensitised her to differences in ways gender plays out. In the early 1970s, she was very involved in the women's movement, which strengthened her commitment to pursue her doctoral work on gender issues. She was hired early on as a 'gender professional' being a Women in Development specialist in 1979–1981, and later a farming systems specialist, an anthropologist, a senior scientist, a community development specialist, a leader of adaptive collaborative management (ACM), and more. Throughout this portfolio of appointments, all served in interdisciplinary contexts, her own purpose and often her remit as well, was to integrate gender. However, as she remarks: 'Not being called a gender specialist probably helped me to be heard by others', briefly hinting at the unhelpful connotations tied to being labelled as one. We asked Carol about her positioning as a 'gender professional'.

Babette: How do you represent yourself to people?

Carol: I was trying to gender mainstream for so many years. It was very easy for me to assess the audience that I was talking to. So, my automatic gut answer if somebody says, 'What are you?' I'd say, 'an anthropologist'. But whether I emphasised the gender part or not, depended very much on the context. If I was talking to Norwegian donors, then I would emphasise gender. If I was talking to a team of forest officers in Zimbabwe, I would emphasise the interest in working collaboratively with local people, and I might stick in there the need to work with women, just to give a little hint that I wasn't only going to be talking to the men. And otherwise, and over time, you become more and more upfront about that in those somewhat partially hostile or potentially hostile contexts.

Carol explains that her background in anthropology has provided more than a set of tools for people-centred forestry work. As a discipline, anthropology enables a holistic perspective for dealing with complex issues. But for the applied anthropologist working in an interdisciplinary setting, there is also a requirement to translate and bridge.

Carol: I am an anthropologist and feel that this discipline has been an excellent grounding for gender studies. But it has also been valuable to work collaboratively with other disciplines, to learn what elements of gender studies 'speak' to practitioners [from other disciplines such as biophysical sciences and economics, for example]. Development-related issues are complex and holistic, and that's what anthropologists investigate. It has been extremely important for understanding the world and finding possibilities for empowerment. This grounding is also important for communicating with other disciplines, although this is one that requires much work as you have to convert the complexities [of social life] into something that fits better with the more linear approach that conventional positivist scientists often prefer. Many of the biophysical scientists I worked with liked to look for *the* [one] cause of phenomena, rather than trying to understand the complex interconnections among parts, including feedbacks and circular and mutual causation. They wanted to simplify to test for one result or another in search of general principles, whereas my own interest was in figuring out the workings of a particular 'whole' that might (or might not) represent a broader category.

Mobilising ethnographic practice to bridge disciplines

Given the debates that have emerged around applied anthropology in development settings, and the challenges other authors have noted with regard to the knowledge hierarchies that can challenge the integration of ethnographic approaches in interdisciplinary contexts dominated by positivist natural and social sciences, we were interested in learning how Carol navigated potential knowledge politics at CIFOR, particularly with regard to work on gender.

Babette: What were the different disciplines at CIFOR? Were there compelling breakthroughs you've experienced in bridging your professional gender knowledge with the technical knowledge of your colleagues?

Carol: CIFOR has an interdisciplinary staff which plans and coordinates research on forest- and people-related topics, with partners in developing countries. There are typically foresters, ecologists, biologists, geographers, anthropologists, sociologists, economists, hydrology specialists, quite a range of disciplines. The mandate of the scientists is to do comparative research mainly that examines the relationships among people, environment and, to a lesser extent, production. Partnership and interdisciplinary collaboration have been keystones of the institution.

The institutional context of CIFOR has favoured an interdisciplinary approach, and much emphasis has been on forest governance and natural resource management through multi-stakeholder work that necessarily requires expertise beyond positivist science. But Carol also notes that a key factor that has facilitated the integration of gender concerns into CIFOR's work has come through a cornerstone of ethnographic practice, that is, fieldwork.

Carol: Going to the field with other people is so, so valuable for making gender differences obvious. Working together in the field is hugely powerful in getting others from a variety of fields to realise that gender is important, but it's not always easy to arrange shared fieldwork (of some duration). It's also useful to see the clever things that women come up with when you give them a chance, as we were able to do in some of the adaptive collaborative management (ACM) projects and CAPRi (Collective Action and Property Rights) work (Colfer, 2005a; Komarudin et al., 2011). This changes researchers' perceptions of women as being too busy or too dumb or too disempowered to act.

An ethnographic sensibility is also put to work in bridging knowledges of colleagues in a very practical sense, and this is another factor that Carol suggests enables a better uptake of gender knowledge in interdisciplinary environment and development work.

Carol: I think we need to tailor our analyses very carefully to the various audiences. We have to meet them where they are: which of course means that we have to double our work, because the transformative work requires sophisticated social scientific analyses whereas the communication with unconvinced folks requires simplification, among other things. I'd say we also need to avoid giving any impression of disdain for the ignorance of our non-social science colleagues (after all, we are as ignorant of their fields). We need to cultivate real respect for others' knowledges, whether villagers or colleagues in research centres or academia. Colleagues tend to be defensive and to fall back on the superior prestige of the 'hard sciences' (which hardens their commitment to positivist approaches) when we behave disrespectfully.

While working in interdisciplinary contexts and doing my gender work, I applied an ethnographic approach to other disciplines. I tried to understand their world view, their scientific traditions and assumptions, their customs. And I think that's been very helpful since I'm able to stand back and think, 'Okay, let's look at their worldview as an agronomist, or as an ecologist, or whatever, from a kind of ethnographic perspective'. I found that it was really helpful working with them more smoothly.

Becky: That's really interesting. So actually, you were using your skill as an anthropologist to enter their lifeworld and language.

Carol: Yes, I came to understand how they look at whatever we're doing together. It helps me. I mean, it's not that we have any kind of big disagreement. I see that it's simply different ways of looking at things. And if I can understand how they're looking at the research situation, then I can rephrase things a bit. This also allows easier interdisciplinary collaboration, I think.

Part of the job of the ethnographic researcher is typically to explore the details of everyday life which otherwise go unnoticed, trying to read the silent scripts that organise ordinary activities (Dove & Kammen, 2015). Ethnographers appreciate the extraordinary-in-the-ordinary that may help to understand the ambiguities and obscurities of social life (Ybema et al., 2009). By doing fieldwork, ethnographers examine the complexities of local contexts and their influence on development outcomes, such as say, in the sustainable forest management project that CIFOR as an organisation aspires to. 'The field' is not only where one observes changing structures and dynamics of local people, but as Carol says, it is also somewhere one can work collaboratively with other disciplines, to learn what elements of gender studies 'speak' to them. It does seem that immersion in field contexts with interdisciplinary teams allows the ethnographer to understand how colleagues, for their part, make sense of their surroundings with the possibility of learning about the social processes that situate their knowledges. This enables her to translate complex social realities into understandable languages and devices that will create common understanding and cultivate in them an appreciation of the social dimensions of things.

Creating common understandings: numbers and generalisations

A perennial challenge for anthropologists working in interdisciplinary development contexts is to develop knowledge that 'works' within the logics of dominant networks within the organisation and/or project in which they are based (Forsyth, 2011): a challenge that is also faced in gender research (Kunz &Prügl, 2019). The use of numbers within participatory research and rapid rural ethnography methods and tools is a well-recognised strategy for bridging across epistemologies through a common research language. Our discussion turned to this issue and the role numbers had played in creating common understandings within Carol's work at CIFOR.

Becky: Could you tell us more how the interdisciplinary collaboration worked out?

Carol: The thing that came to mind immediately was the Who Counts framework that I developed. It's a matrix. It has seven dimensions, and these are dimensions that I developed when I was realising what my forester friends were not noticing, when we were in the field together. The goal of that particular project was to develop criteria and indicators for sustainable forest management. That was the name of the project. And my component of it was on human well-being. Well, none of us really had any idea of how to start looking at that in a forestry context. We were working with very, very marginalised peoples in West Africa. I came up with some things that I thought were really central to involving marginalised people in forest management, and I fashioned those into seven dimensions. I was also aware my companions were completely numbers-oriented. Anything that you could possibly convert into a number would be much more acceptable to them, believable by them. And so, I made this matrix which had the stakeholders across the top, such as the original inhabitants, the migrants in the villages and the forest service, which differed in every place. We listed six or seven across the top, and then we had these seven dimensions down the side, and then we went to various people who knew about the area and we asked them to give a score for each dimension, for example, the severity of poverty. It's nothing precise, but it does kind of give you an idea of the differences among the different stakeholders on these dimensions that are important for human well-being. I'm not a very quantitative person, but I had to think of ways to do simple quantification because it really helps in communicating with non-social scientists. I've become a great believer in triangulation that includes some simple quantitative elements.

The resulting tools were published by CIFOR as a series of guides, similar to the rapid and participatory rural appraisal tools that have been associated with the integration of ethnographic approaches within natural resource management research (Colfer et al., 1999a; 1999b; 1999c). They continue to be used: Carol notes that she has recently used some of the tools in research in forest communities in the western United States.

Babette: What are your starting points in integrating gender with forest resource management? Do you consider impacts of deforestation? Do you look at the paucity of gender-disaggregated data? Do you look at the experiences of women (and men)?

Carol: I have always been concerned with all of those things. My ideal starting point is to work collaboratively with women and men, so we can get a feeling for the holistic context and integrate intersectionality issues. I like to develop collaborative goals and related plans based on what local people want and figure out ways to marry that with wider-ranging goals and concerns from the outside. I think this is the only way we'll get any kind of sustainable solutions (to the changing and variable conditions we encounter in the field). That was possible with the ACM (adaptive collaborative management) project, which I led, but has not been with many of CIFOR's projects (Colfer, 2005b).

Aside from developing tools that involve some simple quantification in order to bridge and triangulate with other knowledge frameworks, we discussed a second issue that is commonly raised in relation to ethnographic work in development contexts, and that is the pressure to scale up and generalise beyond the local context. Here, the principal line of tension is between quantitative (which emphasise the extent of phenomena) and qualitative (the meaning of phenomena) approaches to knowledge.

Babette: Were your technical colleagues concerned about knowledge being biased and the numbers having to be very big for them to actually mean anything?

Carol: The jobs that I've had at CIFOR have all been community-based – either community-based or small area-based. Given that we were working in small-scale contexts, dealing with big numbers has not really been such an issue. We talked a lot at CIFOR about different scales, and it was acceptable to the institution for me to focus on the local level, which provided a framework for the numbers we could come up with. But if I'd been purely qualitative, it would have been kind of alien and they would have worried more about how to assess it.

Babette: As much of your work was actually community-based, was there any concern from within CIFOR or any other organisation you were affiliated with to actually try to find out what can be generalisable from the community-based studies you've done?

Carol: When we first started the project called adaptive collaborative management, I had that problem again and again, both from the administrators within CIFOR and also from the Board of Trustees. And I argued that if we're trying to manage forests, they are in a particular place - something CIFOR scientists readily recognised. Every forest is different, every community is different. We really are going to have to come up with some techniques of finding out about peoples' lives that are transferable. We can't transfer the information specifically about that place. We have to transfer the ability to find out such information about each individual place. And I really believe that.

In other words, what has been scalable are the tools themselves, rather than the knowledge produced by the tools.

Pragmatics of positionality and participation in ethnographic action research on gender

Work on feminist epistemology and research places fundamental importance on how research is conducted (regardless of singular or multiple methods) and to what purpose (Harding, 1987). Reinharz (1992) characterises feminist research as being (i) focused upon analysing and understanding gender within the context of lived experiences; (ii) committed to social change, and (iii) committed to challenging thinking about the researcher's subjectivity and the relationship between the researcher and the researched. The commitment to change and attention to the relationships with research subjects are key to feminist research. Practices of reflexivity – researchers reflecting on their positionality, critically examining the research process, and the commitment to change (Hesse-Biber, 2014), and attention to the relationships between the researcher and the research subjects (Stacey, 1988; Nagar, 2003; Craven & Davis, 2013) – stand out as the defining characteristics of feminist research.

Diane Wolf (1996) additionally draws attention to the uneasy situations in the research process, which she argues, centrally revolve around power: power differences stemming from different positionalities of the researcher and the researched (race, class, nationality, life cycle and age, urban-rural backgrounds); and power exerted during the research process, such as defining the research relationship, unequal exchange, and exploitation. Underlying all these is the feminist imperative to transform and create positive change, which undoubtedly has its share of uneasy arrangements and troublesome outcomes. How have these kinds of issues arisen in Carol's work? Given the pragmatics of applying anthropology in an environment and development context, what kinds of compromises and reversals have been encountered that complicate the 'agency' of the researcher to follow a participatory and transformative agenda that is aspired to in academic discussions, for example?

Babette: How do you see questions of your own positionality affecting your role, practice, authority and voice as a gender professional?

Carol: At the moment, I'm an elderly white-haired woman with lots of experience, so that makes things easier, I'd say (though admittedly I've changed my role to one of basically consulting, rather than being central in the organisation, as I was when I was more energetic and directly involved in all the politics). Being white, speaking English, being American – all have made my life easier, giving me automatic power in contexts where I could have been marginalised (and would have been when I was younger, and also when the world social context was less attuned to gender issues). My own positionality also affects my feelings about gender, the social arrangements I find pleasant, agreeable, desirable (which may not be what those I work with seek). This has sometimes been a dilemma for me. In the Middle East, for instance (before my CIFOR work), I felt some horror at the idea that women needed taking care of; yet Omani women and other Middle Eastern women I met had grown up seeing that as something positive, seeking a man who 'would take care' of them. I think I was able to damp down my dismay, but I always have struggled with trying to do what was right, juggling what I personally desired, with what those I was working with desired for themselves.

Saying all this, I was thinking about my role in CIFOR and the potential for me to be marginalised. But there's another element, which is my role in interaction with villagers. Although I was always aware of power dynamics, the degree to which my own position gave me power was not as clear to me as it became when we moved from village level work to work at the district level. In the latter context, it quickly became clear that district officials in Indonesia and elsewhere, had considerably more power in our relationships than had the villagers I'd dealt with (Colfer, 1983). The officials could refuse to collaborate, contradict or disagree with conclusions we put forth, move key players into other roles; withdraw funding; etc.

Becky: Some of the hesitancy, even resistance to doing work on gender with a transformative agenda, is that we're intervening in other people's cultures. And you've said that they do need fixing in a way. But I wonder if you've got any examples about field contexts where you faced that sort of tension in a very real sense and were caught between doing something and inaction. Was there anything where you stepped back a little bit from something because you felt uncomfortable around some of the transformations that your intervention could then lead to?

Babette: Yes, in fact, disrupting cultures is always made as a popular excuse not to meddle with gender inequalities.

Carol: Yes, people say, 'Why are you doing gender? You're trying to change cultures.' But what about the big logging or plantation industries who come into a community's traditional territory, take over or adversely affect their means of livelihood, influence local level politics? They are also changing gender (see Elmhurst et al., 2016) So, I guess it's just a matter of perspective. I do have a few examples to share where there were some tensions that had to do with disrupting the way things were being done.

The examples Carol outlines below from various projects she worked in also reveal the limits of researcher agency in working collaboratively at community and local government level in contexts where strong gender norms restrict what is achievable in the research.

Carol: In Zimbabwe, as part of CIFOR's ACM program, we were working with women to make their use of local grasses more sustainable – a goal they had identified themselves (Standa-Gunda et al., 2003; Vanclay et al., 2006). They'd developed a new way of making brooms that involved more sustainable use of the grass, as well as being more attractive. They wanted to bring their brooms to a trade fair the government was organising in another town, as a way of disseminating the new technique and also to sell the brooms they'd made. The idea was concocted jointly by the women, our fieldworkers, and government officials. The husbands, fearing danger and infidelity, did not want the women to go to the fair. In discussions it became clear that there would be no problem if the women went in a group, so the trip was planned that way (Mutimukuru-Maravanyika et al., 2008).

Recognising that women were also actively involved in fishing in Danau Sentarum Wildlife Reserve (now National Park) in West Kalimantan, we planned to bring together fishers from each community in the reserve to discuss ways to manage the fishery in a better way (Colfer et al., 1996). We asked that each community send three people, with at least one woman. Very few communities sent their woman representative. We conducted the meeting anyway. At least the idea had been floated.

In the same community, I arranged to accompany a woman to her rice field, which was behind her village. I wanted to be alone with her because I was new in the community and my Melayu language skills were minimal. I found it easier to understand one person at a time; and I also wanted to get the woman's views. My assistant was a man. He and she both agreed to this plan, and I went off alone with her. We were having an excellent time; she was getting progressively more comfortable and willing to talk, when he decided to come and join us – with a whole band of villagers! I did not handle the situation well, stomping off in irritation. He said he'd come to help me with the language; but he may also have come because the community may have been fearful that I'd hurt her. There was a long history of conflict between the neighbouring Iban and these Melayu, with the Iban particularly feared as (ex)headhunters – though there had not been any direct conflict for decades. There remained significant fear of strangers though, with rumours of maltreatment rife. At one point, for instance, there were concerns that we planned to sacrifice a local virgin and plant her under the support beams of the Danau Sentarum Field Center. The woman I was interviewing was genuinely frightened after I stomped off, never to speak with me again. This was not my most stellar fieldwork moment.

Carol's accounts above demonstrate to us that efforts towards being inclusive in field contexts – or the 'feminist will to improve' – are fraught with tensions, usually prompting us to resort to compromises and pragmatic responses that fall short of our original well-intentioned goals. In situations where local patriarchal norms are particularly strong, feminist research interventions often prove to be challenging. We are also reminded of the need to bridge the gap between research agendas and change that's needed on the ground, as it is increasingly becoming clear that even as social research seeks to analyse and address persistent inequalities, marginalisation and disadvantage, new layers of inequality keep emerging, ever more subtle and difficult to deal with. Carol notes that certain structural inequities related to capitalism, or normative ones related to religions are particularly intransigent.

Feminist scholar-activist researchers have stressed the importance of developing research questions and analytical frameworks with communities so that research findings do not become inaccessible and meaningless to the people whose lives the research is intended to improve (Nagar, 2003), an approach that echoes early iterations of participatory research, sometimes dubbed 'rapid rural ethnography' (Chambers, 1994. Both the ACM and CAPRi programs that Carol led were structured specifically to address this issue. Great efforts were made to ensure that the issues addressed, the solutions offered, the ways of monitoring progress, and assessment of results – iterations built into the approach – came from the local people. However, in development research, where Carol and Babette focus their efforts, research agendas are often pre-arranged, guided in large part by the normative goals of the donors and development paradigms in currency. This is illustrated in what Carol describes as a less successful attempt with adaptive collaborative management, when the leader of the broader team was not sufficiently convinced of the need for local level decision-making. He was tethered to the wishes of his donor, which meant the approach simply did not work (Colfer et al., 2011). It is often envisaged that the role of (gender and) development research is 'to inform appropriate and sustainable development policies and practice'. Without a solid understanding of the local context and local needs as well as of broader national, state, civil society pressures, development interventions often fail to deliver *intended* outcomes and can exacerbate existing challenges (Hammett et al., 2014; italics ours). There is a widespread, implicit assumption, therefore, that development agendas are unproblematic and politically uncontested, and take the needs of 'development beneficiaries' fully into account. Applied social sciences in the field of development have also historically resided outside academia where clients expect concrete policy recommendations to respond to *a priori* goals and intentions (Ervin, 2005), which is at odds with both an ethnographic approach and participatory action research. In one sense, the space for agenda-setting is somewhat constricted and can leave gender professionals with minimal room to manoeuvre. However, in another sense, one can rely on and be creatively opportunistic by tapping the underlying ethos of development research, which is to generally effect equitable and sustainable outcomes, a normative engagement that is not always present in 'blue-sky' academic research undertakings.

Politicised intersectionality

Our final area of discussion focused on the incorporation of recent feminist theorising into the applied anthropology context of Carol's work with CIFOR. Given the centrality accorded to social difference in Carol's work, what kind of learning has there been in bringing academic conceptual frameworks into everyday practice? Intersectionality is a key concept that has considerable currency in academic feminist work. It originates as an activist concept, coming from the US black civil rights movement (Crenshaw, 1991). As a way of identifying and acting upon themes of power and marginalisation, it has since been incorporated into academic work as a concept for identifying the coproduction of inequalities through gender, race, class, age and so on (Mollett and Faria, 2013; Nightingale,

2011), with some arguments suggesting it has lost its radical theorisation of white privilege and power on the way. Debates continue, even as the concept is being incorporated in order to add nuance to the gender frameworks used within environment and development organisations, including CIFOR. What has this meant in the context of Carol's work?

Carol: I re-read some stuff I wrote in 1977 after I finished writing an intersectionality manual recently. And I thought, 'God, is there really nothing new? I already wrote about that.' But looking at intersectionality – that's another thing that's difficult to translate into numbers because there's so many different categories. If you're a numbers freak, then you're not going to like intersectionality very well [laughter].

Becky: Too many variables floating around [laughter].

Babette: Talking about intersectionality, it is something that people nowadays read about a lot. And I agree with you that I think it's been around for quite some time, but we just didn't label it as intersectionality. I'm also hearing some feedback that maybe it's better to just use the term intersectionality rather than gender. What do you think about that? Or it's a way of looking at other forms of social differences and power, but at the same time de-centering gender. What do you feel about that?

Carol: Bimbika Sijapati Basnett and I began working with a colleague on the intersectionality manual I just mentioned, a couple of years ago (now published as Colfer et al., 2018). He was young, very smart, just out of graduate school, and he knew a lot about intersectionality. He argued against our approach of beginning with gender. He felt that gender should be presented as just one of the intersectional elements of identity in need of attention. Bimbika and I both felt that gender is something that is in every culture. It transcends everything else, is in all of the identities and none of the other identities is as ubiquitous as gender. We were also working within a program that focused on gender, so that was another more practical consideration. We won the argument but I don't know if we ever convinced him that this was the best way to go or not [laughter].

Becky: In the 'revisiting gender' project, Babette, do you remember, we did have this long conversation and decided that there's always an ethical question about if you don't mention gender it will just vanish completely? It will. It always has. And we need to be bold in making that statement, really.

Babette: Right. If development professionals can do away with it, they will.

Carol: I think many of us have had this struggle. And you can kind of throw some statistics around as well, which I think the UN does, on some of the gender inequities that are still incredibly persistent. Gender problems have not been solved. They keep resurfacing.

I think we would probably be able to reduce some of the antagonism about gender if we did a better job of looking at men in the same way that we're looking at women in gender research. I really feel it's time to do that, and began reading about it a few years back – there is some masculinity literature [as cited in Colfer 2020]. But some of it is so negative about men – it's even written by men, but it's pointing out problems that men create and also suffer. I don't think we've done a very good job of looking at men in the same sympathetic way we look at women analytically. The masculinity issue, particularly, and also in the ways that we've contrasted different femininities.

Becky: Yes, I think that goes missing when there are gender focal points in organisations, and that's often translated as the one who fixes women or sorts out women questions, rather than gender. It could be about masculinities within an organisation as well.

Carol: And masculinities are very, very powerful forces. I really think we need to pay more attention to them. I was thinking about it when I was doing some research in the US in 2017. This was a community where I did three years of ethnographic research in the 1970s; one dominant ideal masculinity for men there was to be a logger. To be strong, to be outdoors, to be in charge, to be bossy to one's wife, to run the household. There was this sort of ideal masculinity that a lot of men couldn't possibly fulfil. And also, when I was young, I remember men were supposed to be the ones who had to support the family; that was their job and they had to do it. Women have now been able to find jobs so that's the other side of the coin. We haven't paid attention to that and we should.

Becky: I'm sure that's right. The burden of being *the* breadwinner.

Carol: When I became a solid feminist in the 70s, I really saw it as a way of opening up avenues for women, but also reducing the pressures on men in this 'breadwinner' role, which was so clear in those days. It's become much less clear now. But there was a lot of pressure on men, and I thought that everybody would benefit. I still do, actually [laughter].

But I think maybe the entry point for this is the growing concern on the youth, isn't it? So applying intersectionality, right? There is now interest in young people in forestry, where masculinities do come in because some of their concerns are about the unemployment of male youth, and what that does in terms of self-esteem and related issues. In my recent US-based work (Colfer, 2018), I've also become sensitised to the differences in American masculinity as people age. It's not entirely clear if the differences I saw in 2017, vis à vis, 1975, were societal differences, age-related differences, or even just differences in my own perceptions! But they were extreme.

Gendered experiences are crucial and fundamental to understanding all things social, as gender is in some way refracted in all social phenomena as is evidenced through the ethnographies produced by Carol, amongst others at CIFOR. Intersectionality has been used not only as an analytical tool, but also as a heuristic to amplify and highlight specific problems that are generally overlooked and silenced (Hancock, 2016). The growing recognition of the importance of intersectionality and its application to new contexts that lie beyond its social justice origins have led to suggestions that its critical edge and transformative potential are being blunted.

This seeming 'depoliticisation' happens when intersectionality is used merely as a descriptive tool without paying attention to its original formulation as an analysis of power dynamics (Mügge et al., 2018; Mollett & Faria, 2013).

Carol: I was thinking that one difference between earlier attentions to intersectional issues and current ones called 'intersectionality' revolved around being less explicit earlier about the power element. Although I did deal with power in my own earlier analyses, I didn't have a concept like intersectionality per se that *required* it, where power was integral to the concept.

The work that has been undertaken by Carol and her team at CIFOR to think through applications of intersectionality in their gender programming has brought internal debates about what this means for 'gender', particularly where efforts to bring 'gender analysis' into institutional practice can feel fragile and temporary. A strategic decision was made in writing the 'manual' to continue to foreground 'gender' but to challenge binary thinking by looking at how gender is produced and experienced through other axes of power and marginalisation (Colfer et al., 2018). There are resonances with Carol's earlier work, but as she notes, the question of 'power' was not necessarily foregrounded even as an ethnographic approach to community-based research in forest contexts begins from a position of complicating simple categories such as 'gender'.

Conclusion

We began this chapter by reflecting on the power of doing ethnography in gender and natural resource management research contexts, and its capacity to understand the plurality of knowledges and with that, bridge diverse groups' understandings of society in interaction with the natural world. In doing our ethnographies, we also encounter uneasy issues of representation of our research subjects, as well as accusations that we are disrupting cultures as we interrogate power relations.

The often-repeated charge of disrupting cultures through gender work is often used misogynistically to undervalue the work itself in order to avoid uncovering or unsettling hierarchical social relations that serve to disadvantage women and other social groups (by caste, class, age or ethnicity). This accusation often ignores real life experiences of women and other groups who may suffer forms of marginalisation. At the same time this ignores the fact that 'culture' is dynamic and changing. Ethnography can be a way of opening up intractable areas of unequal gender relations and can thoughtfully challenge notions that 'culture' and cultural norms represent a 'no-go area' or a dead-end, ideas that disrupt emancipatory processes towards gender equality, rather than advancing it. Bringing social movement concepts such as intersectionality into applied contexts via academia is not always a smooth process, it complicates an emphasis on 'gender', prompting debate as to whether gender should be pre-assumed to be a key axis of power or whether that assumption should be converted into an empirical question: under what circumstances is 'gender' the principal axis of power?

Carol's extensive experience as an anthropologist working in interdisciplinary settings around the world, and the reflexivity that accompanies an ethnographic approach have demonstrated the importance of seeing the world from multiple perspectives: not only from those of the local communities with which she works, but also those of government bureaucrats and her fellow scientists. The advantage of her ethnographic approach comes through its principal mode of data collection: community-based fieldwork. Working in field-based contexts brings to life the gender analyses that might otherwise be unconvincing for some. Her work has involved some multi-directional navigation across the boundary between academic and applied anthropology in her work as a gender professional. Bringing an (academic) intersectional analysis of gendered power into that reflexivity complements her observation that the best laid plans for inclusive community engagement in projects seeking to improve well-being and participation can be derailed by sticky gender norms that can also challenge the agency of the feminist ethnographer in making a difference.

References

Bebbington, A., Guggenheim, S., Olson, E., & Woolcock, M. (2004). Exploring social capital debates at the World Bank. *Journal of Development Studies*, *40*(5), 33–64.

Chambers, R. (1994). The origins and practice of participatory rural appraisal. *World Development*, *22*(7), 953–969.

Colfer, C. J. P. (1983). On communication among 'unequals'. *International Journal of Intercultural Communication*, 7, 263–283. Reprinted in C. J. P. Colfer, M. Elias, B. Sijpati Basnett, & S. S. Hummel (Eds.), *Earthscan Reader on Gender and Forests*, London and New York: Routledge, 2017.

Colfer, C. J. P. (2005a). *The Complex Forest: Communities, Uncertainty, and Adaptive Collaborative Management*. Washington DC: Resources for the Future.

Colfer, C. J. P. (Ed.) (2005b). *The Equitable Forest: Diversity, Community and Resource Management*. Washington, DC: RFF/CIFOR.

Colfer, C. J. P. (2018). 'Before and after the 'Timber Wars': Context, change and potential collaboration on the Olympic Peninsula. *Humboldt Journal of Social Relations*, 40, 137–162.

Colfer, C. J. P. (2020). *Masculinities in Forests: Representations of Diversity*. The Earthscan Forest Library. London: Routledge.

Colfer, C.J., Sijapati Basnett, B., & Ihalainen, M. (2018). Making sense of 'intersectionality': A manual for lovers of people and forests. *CIFOR Occasional Paper*, 184, 40.

Colfer, C. J. P., Andriamampandry, E., Asaha, S., Lyimo, E., Martini, E., Pfund, J. L., & Watts. J. (2011). Participatory action research for catalyzing adaptive management: Analysis of a 'fits and starts' process. *Journal of Environmental Science and Engineering*, 5 (1), 28–43.

Colfer, C. J. P. with Prabhu, R., Gunter, M., McDougall, C., Miyasaka Porro, N., & Porro, R. (1999a). *Who Counts Most? Assessing Human Well-Being in Sustainable Forest Management*. Edited by Ravi Prabhu. 9 vols. Vol. 8, *C&I Toolbox*. Bogor, Indonesia: CIFOR.

Colfer, C. J. P., and others. (1999b). *The BAG (Basic Assessment Guide for Human Well-Being)*. Edited by Ravi Prabhu. Vol. 5, *C&I Toolbox*. Bogor, Indonesia: CIFOR.

Colfer, C. J. P. , Brocklesby, M., Diaw, C., Etuge, P., Gunter, M., Harwell, E|., McDougall, C., Porro, N.M., Porro, R., Prabhu, R., Salim, A., Sardjono, M.A., Tchikangwa, B., Tiani, A.M., Wadley, R.L., Woefel, J., & Wollenberg, E. (1999c). *The Grab Bag: Supplementary Methods for Assessing Human Well-Being*. Edited by Ravi Prabhu. Vol. 6, *C&I Toolbox*. Bogor, Indonesia: CIFOR.

Colfer, C. J. P., Wadley, R. L. & Widjanarti, E. (1996). Using indigenous organizations from West Kalimantan. In D. M. Warren, P. Blunt & D. Marsden (Eds.), *Indigenous Organisations and Development*, (pp. 228–238). London: Kegan Paul International.

Craven, C., & Davis. (2013). Feminist activist ethnography. In C. Craven (Ed.), *Feminist Activist Ethnography: Counterpoints to Neoliberalism in North America* (pp. 1–20). Lanham, Maryland: Lexington Books.

Crenshaw, K. (1991). Mapping the margins: Intersectionality, identity politics, and violence against women of color. *Stanford Law Review*, *43*(6), 1241–1299.

Crewe, E., & Axelby, R. (Eds.) (2013). *Anthropology and Development: Culture, Morality and Politics in a Globalised World.* Cambridge: Cambridge University Press.

Dove, M.R., & Kammen, D. M. (2015). *Science, Society and the Environment: Applying Anthropology and Physics to Sustainability*. London: Routledge.

Elmhirst, R., Siscawati, M., & Colfer, C. J. P. (2016). Revisiting gender and forestry in Long Segar, East Kalimantan: Oil palm and divided aspirations. In C. J. P. Colfer, B. Sijapati Basnett, & M. Elias (Eds.), *Gender and Forestry: Climate Change, Tenure, Value Chains, and Emerging Issues* (pp. 300–317). London: Earthscan/CIFOR.

Ervin, A. M. (2005). *Applied Anthropology: Tools and Perspectives for Contemporary Practice*. Boston, MA: Pearson Education, Inc.

Eversole, R. (2018). *Anthropology for Development: From Theory to Practice*. London: Routledge.

Forsyth, T. (2011). Expertise needs transparency not blind trust: A deliberative approach to integrating science and social participation. *Critical Policy Studies*, *5*(3), 317–322.

Gardner, K., & Lewis, D. (2015). *Anthropology and Development: Challenges for the Twenty-first Century*. London: Pluto Press.

Hancock, A. M. (2016). *Intersectionality: An Intellectual History*. Oxford: Oxford University Press.

Harding, S. (1987). Introduction: Is there a feminist method? In S. Harding (Ed.), *Feminism and Methodology* (pp. 1–14). Indiana University Press.

Hammett, D., Twyman, C., & Graham, M. (2014). *Research and Fieldwork in Development*. London: Routledge.

Hesse-Biber, S. (2014). Feminist research: Exploring, interrogating, and transforming the interconnections of epistemology, methodology, and method. In *Handbook of Feminist Research: Theory and Praxis* (pp. 2–26). Thousand Oaks CA: SAGE Publications.

Komarudin, H, Siagian, Y. L, Colfer, C.J.P. with Neldysavrino, Yentirizal, Syamsuddin & Irawan, D. (2011). The role of collective action in securing property rights for the poor: A case study in Jambi Province, Indonesia. In E. Mwangi,, H. Markelova & R. Meinzen-Dick (Eds.). *Collective Action and Property Rights for Poverty Reduction: Insights from Africa and Asia* (pp. 235–269). Philadelphia: University of Pennsylvania Press.

Kunz, R., & Prügl, E. (2019). Introduction: Gender experts and gender expertise. *European Journal of Politics and Gender*, *2*(1), 3–21. https://doi.org/10.1332/251510819X15471289106077

Lamphere, L. (2004). The convergence of applied, practicing, and public anthropology in the 21st century. *Human Organization*, 431–443.

Li, T. M. (2011). Rendering society technical: Government through community and the ethnographic turn at the World Bank in Indonesia. In D. Mosse (Ed.) *Adventures in Aidland: The Anthropology of Professionals in International Development* (pp. 57–80). Oxford: Berghahn.

Mollett, S., & Faria, C. (2013). Messing with gender in feminist political ecology. *Geoforum*, *45*, 116–125.

Mügge, L., Montoya, C., Emejulu, A., & Weldon, S. L. (2018). Intersectionality and the politics of knowledge production. *European Journal of Politics and Gender*, *1*(1–2), 17–36.

Mutimukuru-Maravanyika, T., Prabhu, R., Matose, F., Nyirenda, R. and Kozanayi, W. (2008). Facilitating adaptive collaborative management in forested landscapes: The Mafungautsi Case Study. In A. Mandondo, R. Prabhu and F. Matose (Eds.), *Coping Amidst Chaos: Studies of Adaptive Collaborative Management from Zimbabwe* (pp. 15–64). Bogor, Indonesia: Center for International Forestry Research.

Nagar, R. (2003). Collaboration across borders: moving beyond positionality. *Singapore Journal of Tropical Geography*, *24*(3), 356–372.

Nightingale, A. J. (2011). Bounding difference: Intersectionality and the material production of gender, caste, class and environment in Nepal. *Geoforum*, *42*(2), 153–162.

Reinharz, S. (1992). *Feminist Methods in Social Research*. Oxford University Press.

Stacey, J. (1988). Can there be a feminist ethnography? *Women's Studies International Forum*, *11*(1), 21–27. https://doi.org/10.1016/0277-5395(88)90004-0

Standa-Gunda, W., Mutimukuru, T., Nyirenda, R., Haggith, M., & Vanclay, J. (2003). Participatory modelling to enhance social learning, collective action and mobilization among users of the Mafungautsi Forest, Zimbabwe. *Small-Scale Forest Economics, Management and Policy*, 2(2), 313–326.

Vanclay, J., Prabhu, R., & Sinclair, F. (2006). *Realizing Community Futures: A Practical Guide to Harnessing Natural Resources*. London: Earthscan

Wolf, D. L. (1996). *Feminist Dilemmas In Fieldwork*. New York: Routledge. Retrieved from https://www.taylorfrancis.com/books/e/9780429962394

Ybema, S., Yanow, D., Wels, H., & Kamsteeg, F. (2009). *Organizational Ethnography: Studying the Complexities of Everyday Life*. London: Sage. https://doi.org/10.4135/9781446278925

10 Disaster risk governance and gender professionals

Command-and-control and re-doing gender

By Bernadette P. Resurrección and Rebecca Elmhirst
In conversation with Maria Holtsberg, Napapan Der Kinderen, and Hilde Jakobsen

The inertia of emergency management in disaster risk reduction

Disasters are now happening more frequently and more ferociously, summoning governments and various stripes of civil society organisations to escalate their actions to ensure more effective protection and safety of lives and property. In the last decades, the growing severity of disasters has beckoned the international disaster response community to more proactively reduce the risks of disasters by addressing their causes rather than limit action to emergency relief and management. In short, this represents a re-framing to pro-active disaster risk reduction (DRR) *before* a hazard can turn into a disaster (Baas, 2008). The disaster anthropologist, Anthony Oliver-Smith (2016), in tracing the evolution of DRR since the 1990s pointed out that DRR was an attempt to frame disaster in terms of risk that was generated by social, political and economic root causes, and among them, development. In 2005, for example, the General Assessment Report of the UNISDR* clearly cited the roots of contemporary disaster risk in historic and contemporary policies and practices of development. What followed was identifying first to third order DRR strategies, where the second and third order ones clearly aimed to neutralise the drivers of vulnerability, exposure and disaster risk.

Humanitarian studies scholar, Dorothea Hilhorst (2003), close to two decades ago, then already lamented that most disaster management organisations continue to pursue a hazard-centric approach akin to emergency management. This meant that they were still focused on delivering relief rather than proactively avoiding disasters by reducing risks, and framing hazards as external to society instead of locating and addressing its social, political and economic causes. Less than 20 years later, Oliver-Smith (2016: 75–76) expresses that nothing much has changed:

> The major institutional focus is still on disaster management and emergency response. The core of the problem is a reluctance to confront the fact that disasters are not external and unforeseen shocks to an allegedly properly

* This has been now changed to UN Disaster Risk Reduction or UNDRR.

> working economic and social system. Unfortunately, in the policy realm risk is still seen as an externality that requires management and preparation rather than as a socially constructed problem created by root or underlying causes that continue to be ignored despite their continuous manifestations in the workings of society. Most resources thus are still focused on emergency management and preparedness aimed toward reducing damage and compensating for losses rather than avoiding risk by addressing the underlying socially embedded causes and risk drivers that generate risk in the first place.

Hilhorst (2003) identified three domains of DRR, they are: (i) science and disaster management; (ii) disaster risk governance, and (iii) local responses. Within the domain of local responses we find a growing volume of grounded gender and disaster research (famously from the throes of hurricanes Andrew, Mitch, Katrina, and cyclones Haiyan and Nagar) that clearly demonstrates that the effects of disasters are profoundly gendered in multiple ways.

Gender and disasters research has developed in response to over five decades of research on emergencies, disasters and catastrophes where a 'calculated blindness' and conspicuous silence around gender persisted (Enarson, 2012: 2). Interest in gender emerged from earlier notions of social vulnerability in disaster contexts, where women and other marginalised groups were found to be disproportionately affected (Wisner et al., 2004; Bankoff et al., 2004). Research has also shown how in many places, women are specifically deprived of protection from natural hazards, which were more available to men and more powerful women in large part due to power relations and social life in highly stratified and conflict-ridden societies (Cupples, 2007; Hyndman, 2008; David & Enarson, 2012). These studies also drew enormous attention to the need to focus on various aspects of gender and its intersections with ethnicity, class, race, age prior to, during and in the aftermath of disasters (Enarson, 1998; Enarson & Morrow, 1998; Fordham, 2011; Fordham et al., 2011; Bradshaw, 2015). However, despite the explosion of research on gender and disasters, DRR practitioners remain frosty to gender-responsive action in their programming. For instance in Bangladesh, when first introducing the concept of gender, DRR staff gave it a decidedly cool reception, saying 'we are here to save lives; we do not have time for gender' (Eklund & Tellier, 2012: 590). The resistance seems plausible, however, other accounts complicate and capture more dynamic interactions between professional staff in the domain of disaster risk governance as will be shown later in the chapter.

Disaster, humanitarian and DRR NGOs are varied, heterogeneous and with different histories. They have become quite complex because of efforts to closely link the longer-term goals of development with humanitarian work with uneven degrees of success and failures. In programmatic terms, organisations may include a diversity of programs that include peace-building, resilience-building, post-disaster reconstruction, disaster preparedness and disaster management. The same group of disaster-affected groups may be approached from a human rights, humanitarian or development perspective. Human rights work addresses legal aspects and protection of refugees; humanitarian efforts address shelter, medical services, food and water; while education and training are important for long-term development and resilience (Roth, 2015: 5).

In her book on humanitarian aid professionals, Silke Roth (2015) explores how women and men re-negotiate their gender positionings as they build their careers in humanitarian aid: by 'doing gender', they invoke and perpetuate gender differences, or by 'undoing gender', they challenge differences between men and women, masculinities and femininities. However, a more counter-intuitive outcome arises for example when women as mothers are made leaders of their organisations. Roth calls this a 're-doing' of gender, or in short, when the most unlikely means to achieve gender equality is through a factor that is believed to fundamentally perpetuate gender inequality.

Roth's study, however, focused on humanitarian aid professionals. Would these same 'doing gender' framings apply for gender experts in DRR organisations?

Despite growing research and attention to gendered disaster outcomes, very little research has pried open the social life of organisations where gender professionals have become part of DRR and disaster preparedness teams. This chapter is an attempt to address this gap through the experiential lens of three gender experts working in this field.

Hilde and Napapan: 'command and control' in the humanitarian sector

Practitioners in humanitarian aid and disaster risk reduction many times crisscross workspaces since there are opportunities today to link these two spheres, both normatively and in practical ways. DRR has been also framed to link more closely with the more long-term goals of sustainable development: 'the rationale for response has shifted from being a matter of public security and safety to one of public interest, investment and safety' (Bradshaw, 2015: 158). In turn, practising DRR is a longer-term process that helps build institutions to be better prepared for, resilient to and able to cope with hazards, and which usually focuses on pre-disaster stages (prevention, mitigation and preparedness), whereas humanitarian aid is usually short-term emergency relief and assistance to those who need help (Baas, 2008). Both Hilde Jakobsen and Napapan der Kinderen work in the DRR sphere but trace their careers to the humanitarian sector.

At the time of the conversation, Hilde was working for a DRR organisation as a senior gender advisor but seconded by a Swedish emergency agency for a specific program on gender and reducing disaster risk in Asia. Earlier, she worked in the humanitarian sector in Africa as a gender-based violence (GBV) adviser. She says she's new to the environment, disaster preparedness, and climate change side of gender work, which is linked more to development. In the program she is currently working in, she brings her experiences from working in refugee camps in Tanzania and focuses on post-disaster human trafficking and GBV.

Napapan der Kinderen works in the same DRR organisation as Hilde and has a background in law, human rights and public health. She says she has not taken up gender studies but has earlier worked on the protection of women and children, as well as GBV, in the humanitarian sector. Like Hilde, she claims she is also quite new in environment, development and disaster risk management. She has been recently hired to serve as the organisation's gender advisor, a position that was a recommendation by a major donor organisation that financially supports the organisation.

Hilde: Our organisation is, strictly speaking, not a humanitarian actor. Humanitarian action usually responds to complex emergencies. DRR work requires more engagement with local actors, supporting governments to be prepared themselves for responding to disasters especially when the international humanitarian community cannot come in. This organisation offers services to governments mainly for building their capacities for resilience-building, disaster risk management programming, and climate change adaptation.

Napapan: Yes, that's right. Our colleagues localise their actions, integrating their work to meet SDG goals and supporting governments, although recent trends in humanitarian work also try to increasingly link with development agendas.

Hilde and Napapan's experiences tell us that the gender issues they worked on while in the humanitarian sector take long for DRR organisations to fully accept and support. Especially since there have been recent efforts to 'localise' both DRR and humanitarian action wherein states can take primary action. Local women's organisations have been largely sidestepped, with far less recognition and support. Meaningful engagement with women and women's rights organisations could increase the success of the localisation agenda (Lafrenière et al., 2019)

Hilde: Napapan, you've been working on reproductive health issues during humanitarian crises. But now in our organisation, these issues do not play a central part and are not that well represented.

Napapan: Yes, I used to work on reproductive health issues in a refugee camp in Thailand. However, here in our organisation, they do not encourage discussion of it with our partners because they fear that that some governments do not welcome these issues. I feel that my competencies are not being fully put to use here.

For example, I was involved in a DRR training for a national disaster management organisation (NDMO) where mostly senior members were present. They rejected the data that I presented about the vulnerability of women being more than men's and questioned me on the source of the data and whether this came from official government records. I have also been told by my senior colleagues to be very cautious about discussing gender issues since many governments do not fully support or agree with these. They tell me stories of how in some trainings, debates became so heated regarding gender-based violence. Some participants simply do not accept that this is happening in their country. This has been a challenge.

Hilde: In humanitarian work, my experience was totally different. Tanzania has been known to have the first ever GBV program in the Burundi refugee camps. This, despite 'violence against women' being taboo in development circles then. So, I think in many ways, the humanitarian machinery can just go ahead and roll out a program, since as they used to say in the camps: 'We run these camps, we're doing this!' They didn't really care what other national colleagues or counterparts thought of the GBV program. And they could do this even without speaking Swahili, the national language. In this specific case, gender issues in humanitarian work have reached much further. And they've done this without really having to connect with the local community. They don't need acceptance on these issues from the national government as they've come in with their international humanitarian machinery. So, the caution that Napapan just described hasn't been so strong in humanitarian programs there since 'they're just going to run and do their own show.' In DRR, you are instead working with governments and so you need much more caution and buy-in, and I suppose the change you make is going to be more sustainable. As I said, despite the GBV program in the refugee camps, there was no GBV program in Tanzania in development circles even 15 years after the camp program as this has been widely thought to be sensitive and taboo. In development work, we have to squarely face and deal with the resistance.

Hilde's experience confirms discussions above about the conventional 'command and control' approach in the humanitarian sector. Ironically, it took this approach to advance the creation of gender programs and may be a process of 're-doing' gender (Roth, 2015), where the most unlikely groups use top-down, undemocratic means to achieve gender equality ends, or at least address the ills of GBV. However, this may have been a unique experience in large part due to the presence of international gender advisors like Hilde in the camps at the right time. Hilde however describes the sector as unapologetically male-dominated, referring to the Swedish emergency agency that seconded her to the DRR program:

Hilde: It's a very masculine domain. Those who are leaders or senior personnel were former fire fighters or are retired military personnel. It's not only a problem of the under-representation of women leaders in many humanitarian agencies, it's also a situation where there are so many incompetent men in leadership. They work for these agencies when their careers are not going well, and they're deployed to be in an NDMO for one or two years. They do not only play the gender card by resisting the presence of women, but they also resist other forms of knowledge other than mechanical engineering. The engineers are seen as the intellectuals among the plumbers and the mechanics. And the social sciences are not considered science.

In DRR as in humanitarian emergency management, strong associations with masculinity still remain, largely because its origins lie with the military and defence institutions of governments. Men still do control the leading emergency management agencies, do still dominate high-status professions and the upper echelons of emergency management organisations and do still benefit as a group from organisational cultures and reward systems tilting their way (Enarson, 2012: 177). Concurrently, many donors continue to conceptualise disaster as natural rather than social events. They continue to focus on their catastrophic outcomes rather than their social, political and economic causes, and as a result, they limit disaster work to catastrophe-related relief and reconstruction activities (Bradshaw, 2015).

Wilson (1999) adds that many of these centres were run by retired military personnel especially under the realm of the fire department, a traditionally male-dominated sphere. For women entering the field of emergency management, women must modify their learned female characteristics to match men's work role characteristics. In the wake of disaster recovery efforts after Hurricane Katrina, Luft (2016: 16) later drew attention to the valorisation of physical labour and a militarised environment that spanned racial groups, which she refers to as racialised disaster patriarchy. Ericson and Mellström (2016: 11) introduce the idea of 'occupational heroism' that also promotes a culture of masculinity: 'Masculinity is institutionalized through networks of artefacts, representations and work practices that are interwoven with and expand certain male bodies.' The humanitarian sector also remains male-dominated at senior management levels (Hoare et al., 2012: 216).

Napapan experienced first-hand the challenges of raising gender awareness among the more conservative corners of governments where hegemonic disaster masculinities continue to thrive (Connell, 2005), and which her organisation was hesitant to unsettle. Both her and Hilde's work shows the uneasy path of gender advocacy that is activated within the continuum between humanitarian emergency management to DRR. Their experience also shows the tensions of combining humanitarian experience with development and DRR-orientated assignments. Experiences in these domains are diverse and sliding from one to the other requires sufficient grounding in the challenges that typically arise from each domain, especially when DRR continues to be framed by the inertia of humanitarian and emergency response programs.

Hilde became pregnant with her second child just before the first year of her secondment to the Asian DRR organisation ended. Her Swedish employers who seconded her to the DRR organisation in Asia did not continue her contract with them because of her pregnancy. She was then forced to leave the organisation because of her pregnancy. This is one more experience of 'command and control' in the disaster and emergency field as told by a gender expert.

Napapan has also already left the organisation as of this writing and has become a free-lance consultant. We now turn to the experience of Maria, showing how the internal social life in a DRR organisation 'does gender' in equally political ways.

Maria: controlling gender in the DRR world

Maria Holtsberg has a background in gender studies and worked for three-and-a-half years as a gender specialist in a DRR organisation which she had just left at the time of this conversation. She has a master's in political science with a focus on women's empowerment and development studies. Earlier, she worked on issues of sexual and reproductive health in a UN agency and a European women's civil society organisation on sexual education. Gender programming, analysis and operationalising have always been components of her various assignments. She made use of her gender background in co-developing large-scale proposals and grant applications in the DRR organisation. However, she was never assigned to lead any program or project on gender, since 'gender' was only considered as something to be inserted in proposals with no standalone institutional presence. Being the only gender expert in her organisation, she began to strategise how to broaden the remit for gender-related activities.

Maria: I started forming a little group of UN and civil society gender experts outside of my organisation and initiating events. I was able to get the experts to do trainings and awareness raising events when my colleagues started hearing more about 'gender stuff' such as GBV from these experts, this gave gender work more credibility and usefulness. However, the lustre from these encounters did not rub off on our internal mundane operations like our recruitment practices, M & E, and our own strategic planning.

In a massive re-structuring process and a new strategic framework that took two years to complete, it was decided that gender would be a core thematic area in Maria's organisation. She then contributed supporting papers to position gender more centrally in the new strategy. As it later turned out, it was decided that gender will no longer be a core thematic area but will instead have to be mainstreamed throughout the organisation. Thus, no leadership position was created specially to oversee gender-related projects and activities, and in the meantime, all new thematic areas – mostly technical in nature – were assigned their respective leaders. All designated leaders were men and were the former heads of earlier departments with new terms of reference that suited and aligned with their earlier scope of work and qualifications. Maria later met with the director of the organisation and he told her that she would lead and oversee the gender mainstreaming committee.

Maria: From my meeting with the director, it dawned on me that this was a junior position being offered to me, although it was one to be directly positioned under his office. My years in this organisation indicated to me that this modality meant that in the end, gender work will not be prioritised. It also dawned more fully on me that gender was not a priority in this organisation. It is also how things work here: the organisation is project-driven, and I did not come with any funding. There is no special position for a gender specialist unless it is donor funded. The proposed new position had no budget attached to it despite being created to serve as an advisory role to the director on gender-related matters. In reality, I also think that traditional and cultural hierarchies and the way they work here will also not enable open consultations between myself and the director. It's just not common practice that seniors consult their junior counterparts in this culture.

The paradoxical situation that Maria found herself in was a case where under the banner of gender mainstreaming, gender work was eventually nailed to the margins. Again, this twist was a case of 're-doing' gender along the terms defined by the organisation: gender ends up everywhere but really, nowhere.

Maria later led the development of a DRR proposal with a strong gender component. She was quite sure that she will not be asked to lead the project even if it resulted from her advocacy, which eventually turned out to be the case.

Maria: Even if the project materialised (which it has), I would not have been rewarded for my efforts, and it would not make a difference. I don't particularly care for the leadership role, but I am more concerned that gender was not given prominence in the overall re-structuring of the organisation and therefore it did not get a leadership position. I think someone with gender competency should be placed in the executive committee level for leverage, otherwise gender being a cross-cutting concern will continuously be side-lined.

What happened, Maria reflects, is also beyond gender, as she saw that employees in the organisation are generally protective of their own specific knowledge spaces and do not wish to expose their lack of knowledge or insecurities for new areas of work. These knowledge spaces constitute the domain of science and management within DRR that Hilhorst (2003) earlier alluded to. In this context, Maria believes that those who were experts in technical fields were assigned the leaders and their knowledge spaces remained virtually protected.

It is also possible that beyond gender lurk uneasy relations between local and international staff like Maria who provides or represents an 'outsider perspective' especially in traditionally 'taboo' areas like gender, which national staff may consider 'foreign' (Roth, 2012). She recalls that the understanding of gender and women's self-organising within the organisation was for lactating mothers to gather together for breastfeeding. This was a case of 'doing gender', marking differences between women and men, but doing so in culturally appropriate and acceptable ways, perpetuating traditional gender roles.

The organisation is also hierarchical where people in management have been long-time employees and young, junior members are given their assignments as a commonly accepted top-down practice. They also socialise and have strong partnerships with national disaster management organisations whose members do not advocate for gender equality as part of their normal order, and as Napapan's experience tells us, resistance is often overt. Some NGOs usually mirror state bureaucracies and adopt a state-centric design in their activities and programs, especially if their corporate identity, core operations and funding support are tied to their role as a service provider to states (Barnett, 2013; Srikandini et al., 2018).

Maria's story reveals that some disaster-related organisations may hire gender experts if it suits their cautionary terms of engagement: no leadership position for any gender expert which then weakens accountability for gender mainstreaming. The core expertise areas are technical services of early warning, forecasting, hazard cartography and GIS [geographic information system], which Barnett (2013) refers to as disaster technocracy. These technical areas were assigned leaders, and they were mostly men.

Finally, these types of expertise also define professional leadership in the organisation because they constitute a hazard-centred paradigm that relies on technologies to control disasters through measurement for prediction purposes (Hilhorst, 2003; Barnett, 2013). Gender would have landed on more fertile and welcoming ground had the orientation been true to the original tenets of disaster risk reduction that seek to address and reduce the gender inequality, social and political drivers of disaster risk in the first place.

The experiences and stories of Hilde, Napapan and Maria tie in closely with the gender and disaster literature about the persistent patriarchal practices in this domain. These patriarchal practices of command and control cut across the geographies of the global North and South, but also build on local intersectional gender, race and expertise politics that render DRR work still difficult terrain for those advocating gender equality.

References

Baas, S. (Ed.) (2008). *Disaster Risk Management Systems Analysis: A Guide Book.* Food and Agriculture Organization of the United Nations.

Bankoff, G., Hilhorst, D., & Frerks, G. (2004). *Mapping Vulnerability: Disasters, Development and People.* Routledge.

Barnett, M. N. (2013). Humanitarian governance. *Annual Review of Political Science, 16*(1), 379–398. https://doi.org/10.1146/annurev-polisci-012512-083711

Bradshaw, S. (2015). *Gender, Development and Disasters.* Edward Elgar Publishing.

Connell, R. W. (2005). Hegemonic masculinity: Rethinking the Concept. *Gender & Society, 19*(6), 829–859. https://doi.org/10.1177/0891243205278639

Cupples, J. (2007). Gender and Hurricane Mitch: Reconstructing subjectivities after disaster. *Disasters, 31*(2), 155–175. https://doi.org/10.1111/j.1467-7717.2007.01002.x

David, E., & Enarson, E. P. (Eds.) (2012). *The Women of Katrina: How Gender, Race, and Class Matter in an American Disaster.* Vanderbilt University Press.

Eklund, L., & Tellier, S. (2012). Gender and international crisis response: Do we have the data, and does it matter? *Disasters, 36*(4), 589–608. https://doi.org/10.1111/j.1467-7717.2012.01276.x

Enarson, E. (1998). Through women's eyes: A gendered research agenda for disaster social science. *Disasters, 22*(2), 157–173. https://doi.org/10.1111/1467-7717.00083

Enarson, E., & Morrow, B. H. (Eds.) (1998). *The Gendered Terrain of Disaster: Through Women's Eyes.* Praeger Publishers.

Enarson, E. P. (2012). *Women Confronting Natural Disaster: From Vulnerability to Resilience.* Lynne Rienner Publishers.

Ericson, M., & Mellström, U. (2016). Firefighters, technology, masculinity in the micro-management of disasters in operational service: Examples from Sweden. In E. P. Enarson & B. Pease (Eds.), *Men, Masculinities and Disaster* (pp. 165–174). Routledge.

Fordham, M. (2011). Gender and disasters. In J. O. Nriagu (Ed.), *Encyclopedia of Environmental Health* (pp. 834–838). Elsevier. http://www.sciencedirect.com/science/article/pii/B978044452272600475X

Fordham, M., Gupta, S., Akerkar, S., & Scharf, M. (2011). *Leading Resilient Development: Grassroots Women's Priorities, Practices and Innovations*. UNDP.

Hilhorst, D. (2003). Responding to disasters: Diversity of bureaucrats, technocrats and local people. *International Journal of Mass Emergencies and Disasters*, *21*(1), 37–55.

Hoare, J., Smyth, I., & Sweetman, C. (2012). Introduction: Post-disaster humanitarian work. *Gender & Development*, *20*(2), 205–217. https://doi.org/10.1080/13552074.2012.698914

Hyndman, J. (2008). Feminism, conflict and disasters in post-tsunami Sri Lanka. *Gender, Technology and Development*, *12*(1), 101–121. https://doi.org/10.1177/097185240701200107

Lafrenière, J., Sweetman, C., & Thylin, T. (2019). Introduction: Gender, humanitarian action and crisis response. *Gender & Development*, *27*(2), 187–201. https://doi.org/10.1080/13552074.2019.1634332

Luft, R. E. (2016). Racialized disaster patriarchy: An intersectional model for understanding disaster ten years after Hurricane Katrina. *Feminist Formations*, *28*(2), 1–26. https://doi.org/10.1353/ff.2016.0023

Oliver-Smith, A. (2016). Disaster risk reduction and applied anthropology: Disaster risk reduction. *Annals of Anthropological Practice*, *40*(1), 73–85. https://doi.org/10.1111/napa.12089

Roth, S. (2012). Professionalisation trends and inequality: Experiences and practices in aid relationships. *Third World Quarterly*, *33*(8), 1459–1474. https://doi.org/10.1080/09700161.2012.698129

Roth, S. (2015). *The Paradoxes of Aid Work: Passionate Professionals*. Routledge – Earthscan.

Sandler, J., & Rao, A. (2012). The elephant in the room and the dragons at the gate: Strategising for gender equality in the 21st century. *Gender & Development*, *20*(3), 547–562. https://doi.org/10.1080/13552074.2012.731741

Srikandini, A. G., Hilhorst, D., & Voorst, R. V. (2018). Disaster risk governance in Indonesia and Myanmar: The practice of co-governance. *Politics and Governance*, *6*(3), 180. https://doi.org/10.17645/pag.v6i3.1598

Wilson, J. (1999). Professionalization and gender in local emergency management. *International Journal of Mass Emergencies and Disasters*, *17*(1), 111–122.

Wisner, B., Blaikie, P., Cannon, T., & Davis, I. (2004). *At Risk: Natural Hazards, People's Vulnerability and Disasters* (Second Edition). Routledge.

11 Lifting the barriers of gender integration in livestock production

By Bernadette P. Resurrección and Rebecca Elmhirst
In conversation with Nicoline de Haan

International research institutes in agriculture such as the centers within the Consultative Group on International Agricultural Research (CGIAR) have historically aimed to produce good science to increase food productivity globally and have re-framed their contemporary goals along the lines of increasing food security and sustainability. The International Livestock Research Institute (ILRI) is a centre within the CGIAR system that was originally set up with the goal of increasing livestock productivity through scientific research as a strategy for reducing poverty. Today, it promotes sustainable and scalable science-based practices to achieve better lives through livestock development by working with both smallholder farmers and commercial livestock producers. It aims to provide persuasive scientific evidence for decision-makers that will enable bigger livestock investments envisaged to deliver better socio-economic, health and environmental dividends. ILRI also recognises women's key productive role in agriculture and in particular, the livestock sector, where they are seen to be often responsible for raising animals and processing and selling their food products (ILRI, 2013).

Nicoline de Haan currently serves as interim director of the GENDER (Generating Evidence and New Directions for Equitable Results) Platform launched in January 2020, operating across 15 centres and aiming to build on the wealth of gender research in the CGIAR system. This position, and its associated program of work, comes about after a long but somewhat uneven development of gender research within the CGIAR (van den Burg, 2019). The CGIAR GENDER Platform was launched in part to raise 'gender knowledge' from being lowest in the ladder of sciences within the CGIAR (van den Burg, 2019: 49) and to offer a vision of 'a world in which gender equality drives transformation towards equitable, sustainable, productive and climate-resilient food systems' (CGIAR Gender Platform, 2019: 6). Recent work has charted the evolution of gender research within the CGIAR, culminating with this new focus (van der Burg, 2019). In this chapter, we develop and nuance this story further with the experiences Nicoline has had in developing gender work within the CGIAR, specifically through her work with ILRI. Given the discursive and historical context of ILRI and the

CGIAR as a whole along the lines of food productivity and sustainability, we wonder how gender experts like Nicoline navigate past and present knowledge hierarchies and the increasingly favourable discursive spaces presently devoted to gender policy. Through our conversations with Nicoline, we reflect on how gender equality and inclusion can be advanced in the work of technical organisations like ILRI, an organisation committed to generate good science on livestock development to accelerate food security through livestock investments. Where do the tensions lie, and what kinds of knowledge coalitions are forged to enable gender equality in the research and policy agenda? We begin with a brief overview of the context of gender research within the CGIAR and ILRI specifically, before turning to the themes that emerged in our discussion, specifically, tackling misunderstandings of gender equality, positions to gender equality, issues in navigating the technical and social interface in multidisciplinary research, the agility of gender expertise, and the implications of building the business case for gender equality in the context of ILRI's work. Our conversation with Nicoline reflects on working within an international food research organisation, a possible site where women's economic inclusion in agriculture and livestock production is seen as an important step on the road to addressing gender equality more broadly.

Positioning gender in the CGIAR and ILRI

The place of gender within the CGIAR has evolved though different discursive pathways. A gender scoping study in 2010 stated that the CGIAR Centres historically have not had a clear gender policy although since the 1980s, there were exceptional examples of gender mainstreaming efforts which, however, were generally not sustained (Kauck et al., 2010). Margreet van der Burg (2019) chronicles the discursive shifts in gender policy and practice over the years up until 2011. International agriculture research began to turn attention to people on fields and farms together with the momentum of the UN Year and Decade for Women (1975–1985). This led to research exploring the differentiation of adoption impacts of new agricultural crops and technologies between farmers groups including women and men, leading to the use of gender to disaggregate research findings. The next wave of gender practices employed a user perspective that targeted women as farmers and natural resource users. CGIAR centres that were more concerned with natural resource management research (forests, land and water) drew from Elinor Ostrom's body of influential work on plural rights, multiple formal and informal institutional arrangements, collective action and polycentric governance among communities in natural resource use settings (Meinzen-Dick et al., 2012; Meinzen-Dick et al., 2014). This led to a new gaze on gendered resource control rights and institutional arrangements that likewise influenced livestock and agricultural research. For example, ILRI and USAID-supported studies have shown that there is need for women to secure and build their livestock assets which they often acquire informally through non-market channels, such as through acts of gift-giving or bequests, but which can also be easily threatened or lost (Kristjanson et al., 2014; Rubin et al., 2010).

Understanding how these informal access and entitlement arrangements shape livestock management is crucial in order to avoid adverse impacts on women who care for and own livestock (Westholm & Ostwald, 2019: 7). Kristjanson et al. (2014) also point out that there is need to increase and sustain women's livestock productivity by improving animal breeds, feeds and health, which are all core to ILRI's scientific work and goals. In addition to the focus on local institutional dynamics and how this provided an entry point for gender research within the CGIAR, the wider application of the farming systems approach began to unpack the farm household, shedding its unitary nature to highlight its differentiated character (van den Burg, 2019).

ILRI (2013) recognises women's key role in food production is central, with recent results estimating that improving women's access to inputs and services has the potential to reduce the number of malnourished people in the world by 100–150 million. This chimes in with the World Bank's view that livestock is primarily an asset for optimising economic well-being and as a social protection measure and economic buffer during stress and shocks (World Bank et al., 2009). Further, 'applying a gender lens to identify and address women and men's different needs and constraints related to livestock production systems and value chains is important for determining the most optimal outcomes as well as the most effective use of resources' (World Bank et al., 2009: 601).

At the time of this conversation, Nicoline guides a team of six researchers at ILRI who explore the linkages between gender and livestock research. She and her team identify entry points to initiate gender-sensitive research that leads to policy action. She works with all technical teams and formulates gender-related research questions, which she sees as the gender mainstreaming part of her work. Nicoline also contributes to the wider gender-related initiatives across the CGIAR centres under the current umbrella of the CGIAR Research Programs (CRPs) for livestock together with other centres such as ICARDA (International Center for Agricultural Research in the Dry Areas) and CIAT (International Center for Tropical Agriculture). The team also conducts standalone gender research, which Nicoline describes as 'strategic research', which she points out to be 'research that we want to do ourselves, how we can position gender more centrally in the livestock research field. Among the themes we pursue is how can livestock empower people. And for this we have funding support.'

Misunderstanding gender and feminism through binary oppositions

Becky: Was there general acceptance of gender in livestock development research at ILRI?

Nicoline: More than five years ago, gender mainstreaming meant that everyone had to integrate gender – the obligations, work and access rights of women and men – in all their work: livestock and animal health, genetics, feed and forage development, institutions and livelihoods, for example. Technical specialists reacted and were questioning why they had to be 'gender mainstreamed'.

They felt threatened by the strong push to bring in gender into their work. They had no clue about how to go about this task, and they did not actually admit that they didn't know. Gender was a top-down thing – although needed then – but it frightened a lot of people. Generally, there is a hesitancy among our technical specialists to deal with social issues, which explains the fear of doing gender work especially in the context of the CGIAR where traditionally, the chief concern and goal was to increase (crop) productivity through scientific means. Gender represents something personal and therefore is strange for many people here.

Babette: Maybe they thought that gender was about being 'anti-men', and of course, this repels people because it's a divisive idea. I hope that this idea has somehow receded?

Nicoline: Years ago, in northern Nigeria we were working on crop and livestock issues focusing on efforts to get manure to be brought to the fields. A team of scientists were meeting and discussing plot sizes to assess how to remove the parasitic weed on plots. I asked if they had asked who owned the plots as this would be important for any follow-up action. Those plots were owned by women and they were not able to get the manure to kill the weeds. It was a powerful moment since I was not a threat anymore to the scientists nor was I the feminist or gender person, but someone who was trying to help scientists do their job better and women with their weeding problems. We were all collectively trying to figure out how to do a good job and make lives better. This avoided the tricky question of empowering women against the men – which makes people uncomfortable.

Popular imagination has persistently retained the oppositional character of feminism and more notably its focus on binary gender relations, splitting roles, interests and actions between women and men, and focusing on recalibrating, evening up between men and women. As Cornwall and Rivas put it (2015: 203)

> the use of the word 'gender' frames two oppositional categories. Like all dichotomies, these are mutually exclusive. Anything that fails to fit the frame is shunted out of it. 'Men' are equated with 'power': 'woman' with powerlessness. 'Men' are the victimisers: 'women' are their victims. Efforts, then, are made to recalibrate these dualisms with talk of 'male responsibility', echoing the dualism in which women are responsible, and men are not; or of 'women's empowerment' in which 'men' are the ones with power and 'women' without. These discourses are premised largely on making good that which is not. They do not offer us the radical reconfiguration of the frame through which social and gender relations are experienced.

Thus, gender mainstreaming and advocacy is also usually tethered to a subordination model, which in framing empowerment as a zero sum game between men and women, could alienate as people may find its missionary and at times adversarial nature deeply uncomfortable (Zalewski, 2010). After having worked as gender advisor in two very distinct development organisations, Ines Smyth reflects on the 'moralism' that sometimes accompanies feminism and gender advocacy in a righteous way, which is defined as 'the pursuit of singular essentialist righteous truth claims about women's lives' (Barriteau, 2003: 132 in Smyth, 2013). As Nicoline's experience shows us, this may frighten others. One way of addressing the limitations of an oppositional binary framing of gender is to draw on feminist post-colonial research literature, which suggests the need to contextualise women as they respond to complex realities and instead depart from the use of reified images of women (often subordinate and victim), without considering how they enter into and engage in social relationships with men within the institutions in their societies (Mohanty, 1988: 77, 78; and Rao, 1991: 11). Experiences in the field, as Nicoline relates, where social complexity is laid bare to demonstrate to technical specialists how technologies such as weeding and manure collection are entangled in gendered resource and property rights, which therefore cannot be sidestepped if technologies are to be adopted. However, another tack is also taken, which rather than challenging the binary frame (as Cornwall and Rivas, 2015 suggest), builds a business case for empowering women in terms of efficiency and raised productivity, which is taken up in a later section. Thus, gender experts fall back on positive messages that associate gender equality with increasing food productivity to gain traction.

The social and technical interface

Babette: I understand that gender research is now more acceptable throughout the CGIAR system. It must be because of years of producing credible gender research in agriculture. Maybe also a lot of intellectual coalescing with technical specialists?

Nicoline: You see, we need to understand where technical specialists are coming from. Generally, they refer to human beings as a research variable. They believe that if you get the technology right, people will adopt the technology. Social scientists will want to know what people want and need and getting them the technologies that would respond to that. I grew up with a father who was a technical specialist, and this made me curious about why they did not ask about people.

Babette: Whereas in the social sciences, we examine how technologies are fundamentally socially created. It must have been strange for technical specialists to think about how gender fits in their work.

Nicoline: Gender mainstreaming seemed a fuzzy idea in earlier years at ILRI. But now we are asking ourselves where and how we can invest on gender and where it can strategically make sense. And this is a change I welcome. What I currently enjoy in my work now is talking with a lot of technical people to try to figure out where and how gender fits in their work. We try to figure out what the gender questions are in each of their themes, such as for instance in animal health. This leads to asking whether we can include women in health care delivery. Or, where and how they can access health care for their livestock. Although I still see that in the past that my work was mainly addressing the main operations in the organisation and leveraging that, we now see an equal balance between gender mainstreaming and standalone strategic and transformative gender research. We had to go through a process with all the technical scientists to see whether gender actually makes sense in their technical areas. By going through this process, they do not need to decide whether gender was important or not in their work, as we help them identify in which areas of their work gender can be relevant. These consultations also enabled me to get to know their work more.

Becky: Could you share with us some examples of how this integration and engagement happens?

Nicoline: Yes, for instance with reference to vaccines: does it really make sense for a vaccinologist to be involved in gender issues? No. But when we talk about the delivery of health care, then it becomes interesting, so we talk to the people who are in charge of that part of livestock health and care. In the past, the focus of efforts was to ensure that women were beneficiaries of our interventions, but we were not looking at what would actually make specific sense for poor women to become beneficiaries. Livestock health and care may make better sense to women who are pastoralists or who take responsibility for them.

There is now constant engagement with technical scientists, and sometimes we want to be a step ahead of them and assess how we can later address gender issues jointly, which I think is the fun part since it's intellectually challenging. We've also reached a point in the work situation where I can tell them: 'that's your work, and this is where I come in'. As a result, there's now more of a conversation about each one's expertise. Which I think is good, because in the past there was a tendency by others who were non-gender specialists to take up the task and say, 'I'll do the gender part.' And we end up with 'bad' gender work. This approach also puts people at ease knowing that we're there to support them, and that they really don't have to do everything. I usually assure them that 'we're not trying to make you gender experts, but we need your help, and then we'll take it from there.' It's not like we're trying to get it into their work, it's just that we're trying to figure out how their research efforts can also help women. I don't see it as us bending it's just that we are trying to develop some options for society. It's actually using technologies to improve everybody's lives.

Reflexive dialogues both with technical colleagues and people on the ground open up opportunities for the inclusion of other voices and bring different knowledges and disciplines into conversation. Reflexivity in a dialogue between gender and technical researchers can also bring to the fore their respective intellectual standpoints and values, which may shape possible collaborative ways forward (Davids et al., 2014). The account above reflects the intellectual boundaries that gender experts often navigate.

These epistemological and normative encounters and interfaces also bring to light what feminist science scholars Evelyn Fox Keller and Helen Longino (1996:

2) have said long ago about the propensity of gender 'to open up an entirely new window on the nature of scientific inquiry.' Nicoline's interactions with technical specialists like vaccinologists, for example, demonstrate how science and technology are often naturalised as self-evident (Björkman, 2005): in the case of producing vaccines, they are taken for granted as a scientific and an interest-free means to boost livestock development. Yet, far from being interest-free, producing vaccines for livestock is intrinsically moored in ILRI's food productivity trajectory described at the beginning of the chapter. Despite this tacit mooring, vaccine production is usually considered a form of neutral, disembodied knowledge until the gender expert, the interlocutor, manufactures an encounter that starts to tie it in with 'what makes sense to women' or what could potentially empower them. 'What makes sense to women' or women's embodied knowledge is usually grounded in their livelihood and personal histories, their gendering process (or how they have become constituted as women-tending-to-livestock), their relational arrangements, and entitlements around raising and caring for livestock together with the animals themselves with their own histories as living beings (Bacchi, 2017; Whatmore, 2002). Vaccines, formerly assumed to be disentangled from social arrangements, visibly acquire a social placement in this interface. The role of the gender expert in these encounters between science and society is that of interlocutor, demonstrating that science, nature and society are intimately intertwined and co-constituted. Feminist science studies have also argued that research has reality-producing effects (Trojer, 2017), since science does not stand apart from the world but actively intervenes in it, re-ordering relationships between humans and non-human others and in this case, re-positioning gender along the elements in the food chain (Greenhough, 2009). Technologies and knowledge also transform when they are adopted, learned and passed on (Björkman, 2005).

Babette: I wonder whether gender specialists like you need to know and be familiar with technical domains and systems.

Nicoline: It is important to be able to carry a conversation, and so for this one needs to understand the technical aspects of livestock breeding as a gender specialist. I was trained as a rural sociologist at Wageningen University, an agricultural university in the Netherlands, so I am somewhat knowledgeable about some of these technical areas. Because of this, I am able to have conversations that I might otherwise have been constrained to have. When we hire people for gender-related positions, I always investigate whether they have a background or understanding about agriculture, otherwise those without will only speak about gender in a vacuum, they may be too theoretical and in the end, this may not be useful for the work that we're doing here. In fact, I moved from the International Water Management Institute (IWMI) because my technical knowledge of livestock is so much better than water systems. I also felt that in the water discussions at IWMI, there was little mention of people.

On one hand, reflexive dialogues provide specialists from diverse backgrounds opportunities for creating epistemological interfaces. On the other, gender experts at ILRI would have fluency in the technical dimensions of livestock management. By knowing these aspects, they then proceed to identify 'entry points' for gender integration and this creates complementary roles for the gender expert and technical specialist, especially if they share the same development goals. This draws attention to the 'agility' of gender both as a concept and as a specific expertise in its capacity to embed itself in diverse, even technical, contexts.

The agility of the gender expert

Becky: Do you think gender work is something anyone can take up?

Babette: That's right, you said earlier that the work of non-gender specialists proved to be counterproductive. I recall that in many places this is done in the extreme: as long as you're a woman, it is assumed that you can do gender-related work.

Nicoline: I would be cautious in hiring someone for a gender position. In the past when the practice was to require all staff to include 'gender' in their work, some people with more technical backgrounds became 'gender literate' but in a way that was unhelpful and generally resulted in unsatisfactory gender research. For hiring gender specialists, I would ask whether the applicant really understood gender, had the theoretical background for it, and understood the issue of power. Those with technical backgrounds, unless they've been trained in gender studies, usually fall apart, and they end up not doing the job very well, which does not help in the long run. For example, when I was relatively new in the job years ago, we were starting to conduct participatory research. I was pleased that my technical colleagues were willing to get involved. So, we trained them and allowed them to do participatory research. Later, I observed that for them, applying the participatory approach meant simply asking farmers questions, which was not fully what participatory research was all about. The results were far from useful, as being participatory requires that farmers and specialists look at problems and solutions together. I am similarly concerned that the same will happen with gender research if we're not careful about who does it. I'm happy to encourage technical specialists, but also think that caution should be exercised if we are to move the gender equality agenda forward. One needs to understand the theoretical foundations of gender.

Research on gender experts has shown that gender expertise by itself is elusive, and is not bound by any one academic discipline, and may be defined by various feminisms (Hoard, 2015; Kunz et al., 2019). However, from her research, Hoard (2015) suggests that the core element in gender expertise is feminist knowledge. Interestingly, Çağlar (2010) further underscores the practice of combining specialist feminist knowledge with other types of expertise in order to engender policy. It is the same with Nicoline who requires fluency in technical agricultural knowledge and feminist knowledge of potential recruits. This draws attention to the agility of 'gender' as a concept and its capacity to embed itself in diverse contexts, negating the idea of a 'universal knowledge' of gender. Gender is 'messy\, both multi-dimensional and intersectional through different axes of power and identity, but also firmly context-driven and socially embedded. In effect, gender expertise is transversal knowledge that crosses disciplinary boundaries and is also contingent on institutional requirements, culture and political circumstances. Harding (2011) further suggests that like race and class, gender is both an analytical and descriptive term, in which as an analytical term, it can frame and explain its diverse manifestations in social relations in different contexts.

Babette: But there are all sorts of responsibilities tied to being a gender expert in an organisation. And we're often tapped for additional services. Nicoline, you did say that there were moments that weren't very comfortable when you were being referred to as a gender professional.

Nicoline: I am uncomfortable when people are talking about being a 'gender person' in a Human Resources (HR) sense, I'll be honest. Which is interesting because there are links between gender in the research and gender in the workplace. That's when I'm uncomfortable. People think because you're the 'gender' expert in research that you're also an HR person and can deal with sexual harassment in the workplace for instance. Having said that, in a lot of cases you do have more theoretical grounding on such issues than a lot of other people. But, still, it's very uncomfortable. It's also very uncomfortable when people use it as a way of pushing responsibility away from themselves. And, I've experienced this where I sense that people are fearful about taking responsibility for looking at gender issues because they're worried about getting it wrong in my presence. I think it would be more useful if we just were at some level considered as social scientists. We will be able to do more because then gender expertise just becomes one of the many identifiers and then people wouldn't be as hesitant to engage.

Gender experts are different from other professions in the sense that there are more explicit ethical and political dimensions attached to their work, largely addressing the general and unsettling goal of transforming unequal power relations and gender inequality in society, including the workplace. This becomes a point of tension especially when organisations declare 'in-house' gender expertise. Prügl (2015) remarks that gender equality is rapidly becoming a valuable goal for organisations and firms because it facilitates achieving all kinds of social goods, including better company management, which creates a good reputation. Strikingly, part of achieving social goods is the growing interest in women's inclusion in economic growth agendas that involve the private sector.

Empowerment by economic means

Babette: How does the current focus on the private sector and including women in value chains relate with the goals of empowerment and gender equality? Your experiences and involvement in these dynamics can probably shed light on this relationship.

Nicoline: For land you need a title deed and usually for a lot of businesses you need actually some title and you need some collateral. Well with livestock, it's there and no deed is needed. You have it, you already have it often so that's a simple one. You can take it with you if you get divorced, or if you're in a conflict area. It self-perpetuates and I call it an ATM. It gives you some money constantly, either through milk or through eggs. And yes, it doesn't mean that you're a big farmer with 7,000 chickens straight away but it gives you some independence, so it gives you some empowerment. It gives you some decision-making power. It gives you that power to make the decision about whether you actually invest in a child's education or not. All those things which give people dignity and it gives them a sense of self-worth. So, I do think that raising livestock makes business sense. We also talk now a lot about what we call the livestock ladder. This is how we can help women who have a few chickens build up to having more. So, if you had a few chickens how can you build up to a few goats? From a few goats, how can you build up to cattle if you would want. Or how can you use that to buy a sewing machine, a milling machine, whatever you want, a chopper for your forages? There are some very interesting things you can do with livestock. With regards to the private sector, I do think we're being pushed more and more into that. But I think let's figure out how we can co-opt it and how we can actually address what women want. I think we often talk about how important it is for women to have their choice. But not all women want to work more. One of the big areas of research we're actually doing right now is also looking at what point does it become a business. But those are also very interesting things especially when you start talking about business and entrepreneurship.

Becky: Actually, some of what you said is somewhat linked to a conversation we had with one of the other contributors in the book about working within the productivity paradigm. How do you demonstrate that there are wins all around if you deal with gender because it will actually enable people to meet many other end goals?

Nicoline: One of the things I do want to say to that is, it was a paradigm I knew when I got into it. I mean, I don't think we're victims of this. No, I knew that, but I wanted to work within the system, and I think we can do a lot and leverage a lot on economic empowerment for women within agriculture. We have to have this bigger end goal of gender equality, yes, but that takes little baby steps.

The discourse of women's economic empowerment has increased currency among a number of influential international development organisations. Women's economic empowerment is considered a requirement for sustainable development and pro-poor growth (GENDERNET, 2011), and is often manifested by women's involvement in market activities and their improved educational attainment. Underlying this principle is a belief that there is a win–win relationship between gender equality and economic growth. The research of Kabeer and Natali (2013) tells us that only certain dimensions of gender equality can positively promote economic growth such as gender parity in education. Whereas the converse, the prospect of economic growth leading to gender equality, is probably unlikely. This is in large part due to the complex and multiple drivers of gender inequality which cannot be mitigated by economic benefits alone. Perhaps enabling economic gains are the little baby steps that need to be taken, but as Kabeer and Natali's research demonstrates, economic gains by themselves do not fully define equality and empowerment.

Conclusion

We saw that there was fear among technical specialists with the prospect of integrating gender issues with their work. We surmise that this is because gender advocacy has been traditionally associated with a sense of moralism that seeks to reduce women's subordination that sometimes elicits contentious reactions. However, over time, and by instigating constructive and exploratory dialogues that provide occasions for reflexivity among small groups of gender, social and technical researchers, a few epistemological interfaces began to create constructive socio-technical *modus operandi*. Gender experts integrate gender issues with science and technology research as they serve as interlocutors linking science and society, dispelling myths that science stands apart as disembodied knowledge. This process of interlocution has also inadvertently reduced the fear of gender integration among technical specialists.

As a 'traveling rationality' (Mosse, 2011), the discourse of women's economic empowerment also travels and circulates widely across different types of organisations, embedding itself as well in research organisations with mandates to address food insecurity and productivity through science. It will be worthwhile to further investigate the ways in which 'women's empowerment' is being integrated into neoliberal economic growth logics and imperatives that have become core to the mandates of contemporary agricultural research institutes and their outcomes on farm communities. As Elisabeth Prügl (2015) suggests in her work on neoliberalism and feminism, it may be necessary to assess what is lost in the process and perhaps what is gained.

References

Bacchi, C. (2017). Policies as gendering practices: Re-viewing categorical distinctions. *Journal of Women, Politics & Policy*, *38*(1), 20–41. https://doi.org/10.1080/1554477X.2016.1198207

Barriteau, V. E. (2003). Confronting power and politics: A feminist theorizing of gender in Commonwealth Caribbean Societies. *Meridians*, *3*(2), 57–92.

Björkman, C. (2005). Feminist research and computer science: Starting a dialogue. *Journal of Information, Communication and Ethics in Society*, *3*(4), 179–188. https://doi.org/10.1108/14779960580000271

Çağlar, G. (2010). Multiple meanings of gender budgeting: Gender knowledge and economic knowledge in the World Bank and UNDP. In *Gender Knowledge and*

Knowledge Networks in International Political Economy (pp. 55–74). Nomos Verlagsgesellschaft mbH & Co. KG.

CGIAR Gender Platform. (2019). *CGIAR Gender Equality in Food Systems Research Platform*. https://cgspace.cgiar.org/handle/10568/107310

Cornwall, A., & Rivas, A.M. (2015). From 'gender equality and 'women's empowerment' to global justice: Reclaiming a transformative agenda for gender and development. *Third World Quarterly*, *36*(2), 396–415. https://doi.org/10.1080/01436597.2015.1013341

Davids, T., van Driel, F., & Parren, F. (2014). Feminist change revisited: Gender mainstreaming as slow revolution. *Journal of International Development*, *26*(3), 396–408. https://doi.org/10.1002/jid.2945

GENDERNET. (2011). *Women's Economic Empowerment*. Issues Paper. DAC Network on Gender Equality (GENDERNET). https://www.oecd.org/dac/gender-development/47561694.pdf

Greenhough, B. (2009). Social studies of scientific knowledge. In *International Encyclopedia of Human Geography* (pp. 204–211). Elsevier.

Harding, S. G. (Ed.). (2011). Introduction. Beyond postcolonial theory: Two undertheorized perspectives on science and technology. In *The Postcolonial Science and Technology Studies Reader*. Duke University Press.

Hoard, S. (2015). *Gender Expertise in Public Policy Towards a Theory of Policy Success*. Palgrave Macmillan.

ILRI. (2013). *Livestock Research for Food Security and Poverty Reduction: ILRI Strategy Summary*. ILRI. https://hdl.handle.net/10568/29037

Kabeer, N., & Natali, L. (2013). Gender equality and economic growth: Is there a win-win. *IDS Working Papers*, 2013(417), 1–58. https://doi.org/10.1111/j.2040-0209.2013.00417.x

Kauck, D., Paruzzolo, S., & Schulte, J. (2010). *CGIAR Gender Scoping Study*. International Center for Research on Women (ICRW).

Keller, E. F., & Longino, H. E. (Eds.). (1996). *Feminism and Science* (Revised Edition). Oxford University Press.

Kristjanson, P., Waters-Bayer, A., Johnson, N., Tipilda, A., Njuki, J., Baltenweck, I., Grace, D., & MacMillan, S. (2014). Livestock and women's livelihoods. In A. R. Quisumbing, R. Meinzen-Dick, T. L. Raney, A. Croppenstedt, J. A. Behrman, & A. Peterman (Eds.), *Gender in Agriculture* (pp. 209–233). Springer Netherlands. https://doi.org/10.1007/978-94-017-8616-4_9

Kunz, R., Prügl, E., & Thompson, H. (2019). Gender expertise in global governance: Contesting the boundaries of a field. *European Journal of Politics and Gender*, *2*(1), 23–40. https://doi.org/10.1332/251510819X15471289106112

Meinzen-Dick, R., Kovarik, C., & Quisumbing, A. R. (2014). Gender and sustainability. *Annual Review of Environment and Resources*, *39*, 29–55. https://doi.org/10.1146/annurev-environ-101813-013240

Meinzen-Dick, R. S., van Koppen, B., Behrman, J., Karelina, Z., Akamandisa, V., Hope, L., & Wielgosz, B. (2012). Putting gender on the map: Methods for mapping gendered farm management systems in sub-Saharan Africa. IFPRI Discussion Paper. https://www.ifpri.org/publication/putting-gender-map-methods-mapping-gendered-farm-management-systems-sub-saharan-africa

Mohanty, C. T. (1988). Under western eyes: Feminist scholarship and colonial discourses. *Feminist Review*, *30*.

Mosse, D. (2011). *Adventures in Aidland: The Anthropology of Professionals in International Development*. Berghahn Books.

Prügl, E. (2015). Neoliberalising feminism. *New Political Economy*, *20*(4), 614–631. https://doi.org/10.1080/13563467.2014.951614

Rao, B. (1991). *Dominant Constructions of Women and Nature in the Social Science Literature* (CES/CNS Pamphlet 2). University of California.

Rubin, D., Tezera, S., & Caldwell, L. (2010). A calf, a house, a business of one's own: Microcredit, asset accumulation, and economic empowerment in GL CRSP projects in Ethiopia and Ghana. Global Livestock CRSp.

Smyth, I. (2013). Values and systems: Gender equality work in different organizational settings. In R. Eyben & L. Turquet (Eds.), *Feminists in Development Organizations: Change from the Margins* (pp. 127–144). Practical Action Publishing.

Trojer, L. (2017). *Sharing Fragile Future Feminist Technoscience in Contexts of Implication*. Makerere University Press.

van der Burg, M. (2019). 'Change in the making.' 1970s and 1980s building stones to gender integration in CGIAR agricultural research. In C. E. Sachs (Ed.), *Gender, Agriculture and Agrarian Transformations: Changing Relations in Africa, Latin America and Asia* (1st ed., pp. 35–52). Routledge. https://doi.org/10.4324/9780429427381

Westholm, L., & Ostwald, M. (2019). Food production and gender relations in multifunctional landscapes: A literature review. *Agroforestry Systems*. https://doi.org/10.1007/s10457-019-00397-1

Whatmore, S. (2002). *Hybrid Geographies: Natures, Cultures, Spaces*. Sage.

World Bank, Food and Agriculture Organization of the United Nations, & International Fund for Agricultural Development (Eds.). (2009). *Gender in Agriculture Sourcebook*. World Bank.

Zalewski, M. (2010). 'I don't even know what gender is': A discussion of the connections between gender, gender mainstreaming and feminist theory. *Review of International Studies*, *36*(1), 3–27.

12 We build the *power* in em*power*ment

Feminist activism at the forefront of environment and climate change arenas

By Bernadette P. Resurrección and Rebecca Elmhirst
In conversation with Kate Lappin

Feminists, not gender professionals

'We are feminist activists', replied Kate Lappin, when asked whether the organisation that she led (at the time of the conversation), the Asia Pacific Forum on Women, Law and Development (APWLD), was an organisation of gender professionals. 'We see ourselves as responsive to a movement and are committed to building power and confronting power.'

The women's movement has grown in size, shape and breadth over many decades, and whenever there has been a need to describe and categorise its membership, both diversity and hybridity often stand out. Notwithstanding debates over the linkage between women's, queer and feminist movements, a core feature of the women's movement can be said to be fundamentally anchored in the political goal to emancipate women from their long history of marginal political, economic and social status across geographical divides and to hence work towards achieving empowerment and equality. Our conversation focuses on what Kate refers to as feminism for the '99%' of humanity, with an emphasis on tackling injustice: shifting the global power of the 1%, which she points out is something quite different from instrumentalised forms of gender mainstreaming.

Gender, environment and development as a field of research, policy practice and activism accommodates varieties of feminism, often differing in their framings of what ought to be transformed and the forms of transformative action. This chapter in particular sheds light on efforts to re-politicise gender along the lines of movement-building that counterpose conventional and technocratic approaches of gender mainstreaming. The ontology of feminist action is to work towards structural transformation by building and strengthening social and feminist movements. The object of transformative action is directed at elite and corporate capture of global resources, the disproportionate accumulation of wealth, and the response is to build movements that challenge these and enable a climate of equal rights, social justice and empowerment especially for those adversely disadvantaged.

Pathways to achieve empowerment and equality, and to engage in movement work diverge for feminists, and often this depends on personal trajectories, social contexts and histories. Our conversation with Kate includes an exploration of how her approach reflects early experiences with labour injustice, women's rights, health and the environment in her native Australia. Her work with APWLD provides an opportunity for us to explore a politicised engagement with 'gender' in the realm of environment and development, specifically in relation with the rights of women affected by extractive industries, and later, climate change.

Kate's parents worked in the state-owned, brown coal-powered electricity industry in a regional town built around the coal industry in Australia. Her mother was dismissed from work once married as government policy prohibited married women from working in public utilities. And so with limited options, her mother became a domestic worker and went on to have six children. Kate started working at 14, and although her early experiences with the Trade Union Movement were not positive due to underlying sexism, she went on to focus her union work on challenging injustice and in fostering women's human rights. At 20, she became the only woman trade union delegate in her workplace and later entered university where she studied in a law school that included a feminist legal studies stream and women's studies. Environmental justice questions greatly interested her because of her family's experience in the coal industry. 'Those early stages of my life influenced me to take up the issues of labour injustice, women's rights and environmental justice. The experience of the La Trobe Valley, the coal dependent town where my parents met, is illustrative of so many intersecting problems. First, the exclusion of women from work once married and the obvious economic independence that led to, then privatisation discarded thousands of jobs and the whole region suffered from economic depression and both the environmental problems locally and globally from the mine got worse with privatisation. Resistance to shutting the power stations is understandably strong in the community and when one of the oldest, dirtiest power stations was finally shut down, the state government supported a just transition package for the remaining workers. But there had been no just transition for my mum or other women denied jobs, no just transition for the workers discarded by privatisation. So, I probably noticed the impact of burning coal from an early age', recounts Kate. She was APWLD's executive director for eight years until mid-2018 and is now the regional secretary for the Asia-Pacific region of Public Services International (PSI), a global trade union federation representing 30 million workers.

Social movements and civil society organisations are known to push political boundaries to pave the way for more open engagements, blaze new discursive paths, and have been chiefly responsible for establishing feminism as a social discourse and policy norm within environmental and sustainability debates (for a complementary discussion, see Chapter 13 on conversations with members of Sida). For instance, the process adopted by the United Nations for formulating the post-2015 Development Agenda has been credited with opening up spaces for civil society and social movement consultations that became germane to the formulation of the Sustainable Development Goals (SDGs). This dramatically altered the rules of agenda-setting in the UN, and which was in large part driven by global civil society's frosty reaction to the SDGs' precursor, the Millennium Development Goals (MDGs), which had been devised with little consultative action. Not everybody was pleased, however, with the outcomes of the SDGs either. Some remain 'cautiously optimistic' (Esquivel, 2016; Esquivel & Sweetman, 2016; Fukuda-Parr, 2016; Razavi, 2016), whereas the creation of the goals left many activist quarters unconvinced of the goals' capacity for meaningfully transforming gender equality. Feminist scholars refer to the underlying propensity of the SDGs to favour neoliberal market-driven means to achieve sustainable development instead of prioritising more redistributive and social welfare measures (Weber, 2017; Koehler, 2016; O'Manique & Fourie, 2016).

Babette: In the gender and environment panel at SEI's Science Forum last 2017 in Bangkok, Kate mentioned that the Beijing Platform for Action, with its emphasis on rights and justice, continues to carry more transformative potential than the SDGs, despite the latter's stated commitment to 'leave no one behind'.

'Yes, I wrote an article in *The Guardian* contrasting the Beijing Platform and the SDGs.* The Beijing Platform acknowledges the structural drivers of inequality including austerity measures, for example. The directives for governments include redistributing military spending to public services and conducting macro-economic reviews to eliminate practices that magnify gender inequalities which would clearly include privatisation. But the SDGs effectively encourage privatisation through public–private partnerships. The Beijing Platform recognises the gendered harms of structural adjustment and directs IFIs to reform but the SDGs require them to stay 'within mandate' – a mandate the banks say is not about environmental protections or human rights.

Becky: How do you locate your position within APWLD? Do you see yourself as a career feminist or gender expert?

'For us at APWLD, the driving motivation is not to build a career. We are not career feminists or gender professionals, and since we are a membership-driven organisation, there is a sense that this is not what our members really want', says Kate, nailing the door shut to being considered a 'gender expert'. 'We also hesitate to call ourselves an NGO, because there's a history of NGOs creating a profession out of other people's misery. The idea of an NGO comes with a particular, perhaps colonial perspective. We see ourselves as a movement, not a technical response to poverty or to climate change. We don't see our work as building expertise, but building power, this is the power in the over used term empowerment. So, we are activists not gender professionals.'

* https://www.theguardian.com/global-development/2015/sep/23/global-goals-womens-rights-pale-imitation-beijing-declaration-and-platform-for-action

NGOs in the Global South present a more complex picture than is often assumed. On one hand, they are essentialised as representing bottom-up commitments to gender equality, poverty reduction and/or social justice. Yet on the other, they may be purveyors of globalised agendas – new forms of colonialism – and elite capture. A conversation between feminist scholars Richa Nagar and Saraswati Raju (2003) explores whether NGOs in India, for example, usher in these new forms of colonialism as they professionalise and conduct gender mainstreaming, leading them to distance themselves from the world of 'real' struggles and align more closely with state and international technical agendas. Raju points out that increased professionalisation means that many NGOs cannot find the time to discuss questions of hierarchy and structure in their own organisations. Adds Nagar: 'The new implicit understanding of the over-arching goal these days is not to overthrow the current system and build a new one', but that 'we have to live with the current political and economic structures and so our aim is to make the poorest women cope better with this reality of shrinking resources and increasing social and economic inequity and injustice by making them more knowledgeable, more comfortable with market forces.' (Nagar & Raju, 2003: 3). The work of Narayanaswamy (2014, 2016) instructs us that 'Southern women's NGOs' are not to be treated as an aggregate or homogeneous category, but instead to recognise that they, for their part, share trajectories of historical and contemporary oppression that allow them to transcend other axes of difference and generate their own elite elements. Nagar & Raju (2003) also note that reports that high-level NGO officials generate are primarily for funders, which De Jong (2016) affirms as she remarks that NGOs are more accountable to their donors who generally encourage service delivery over political advocacy.

APWLD is a regional women's network in the Asia-Pacific region that started as a law- and human rights-related alliance of national women's organisations and is membership-led. It now focuses on movement-building working with rural, indigenous and urban poor women especially in vulnerable climate change contexts. Says Kate: 'When you work with grassroots women, they are more likely to be interested in livelihood issues around their daily lives than perhaps legal frameworks, which APWLD often started with in its past.' Through a research and women's self-organising program they call 'feminist participatory action research' or FPAR, local and grassroots women acquire research skills to understand their own contexts, self-organise and build their own movements to seek solutions for their own problematic conditions. 'We've used FPAR in a palm oil displaced community. What would a feminist approach mean for women by way of building their own power and movement, for example? How could they contest the displacement?' she adds. These are the concerns that their work addresses, which are different from other types of interventions such as income generation and creating local groups for resource management.

Kate says that they also engage internationally and regionally around climate and environmental issues, 'so that we can help create just standards. Our work has overall increased especially since we are also addressing the impacts of the extractive industries on rural women apart from climate change,' she adds. What began as largely a loose network of lawyers and academics, APWLD is now responsive to environmental issues of currency and has become a presence in international meetings such as the Women's Major Group, a group of civil society organisations that engaged intensively in the formulation of the Sustainable Development Goals and the Women and Gender constituency of the UNFCCC. This explains why the network has grown in breadth and is active in multiple scales of political engagement. 'If we need to build expertise, we use the expertise for the benefit of our members', remarks Kate, 'the driving motivation is not to build a career, that's why we don't call ourselves "gender professionals"'. For Kate and the members of the 30-year old feminist network in the Asia Pacific region, the challenge is to address and transform the 'nexus of patriarchy with militarism, fundamentalisms and neo-liberal economic globalization' that undermine women's human rights and freedoms (Asia Pacific Forum for Women, Law & Development (APWLD), n.d.).

The task that Kate and APWLD members have set for themselves, however, is far from straightforward. In a climate where certain brands of feminism have increasingly become *de rigeur*, it may be difficult for Kate and allies to separate the chaff from the grain.

'Doing gender' in a safe way

Many organisations understand gender mainstreaming as establishing and maintaining gender balance, or the equal representation of women and men. Says Kate: 'Many organisations are far more pragmatic and would be satisfied to just get 'gender' in to UN text. For example, with respect to the gender resolutions at the UNFCCC such as the creation of the Gender Action Plan (in 2017), some organisations within the Women and Gender Constituency were just intent to get a commitment to equal representation of women and men in delegations. At that time, some gender experts were keen to get a resolution at any cost without addressing fundamental issues such as loss and damage, financing, addressing injustices. That's 'doing gender' in a safe way.

Bacchi and Eveline (2010) point to similar tendencies in gender mainstreaming that privileges an *a priori* recognition of 'sexual/gender difference' supporting the ontological view that assigns biological characteristics to specific human beings thus setting them apart and unequal to each other. This approach underpins gender mainstreaming efforts that are plainly satisfied with gender parity in representation, as Kate's example above shows. This act dissuades from actually focusing on the politics that privilege some and not others. The focus on differences (often binary differences between women and men) fails to recognise that the acts of allocating or claiming difference is fundamentally politically imbued, and which therefore transformative change. As Fiona transformative change. As Fiona Wilson (1996: 834) argues, 'instead of looking at gender as a difference perhaps we need to look ... at how this is done'. To see how gender is 'done', Kabeer adds: 'we need to analyse the ways in which gender is constructed as a relationship of inequality by the rules and practices of different institutions' (Kabeer, 1994" 84). A feminist political ecology approach locates the political production of gender in the competition of resources that create and stratify multiple and intersectional subjectivities; whereas the gender difference approach, which falls under the suite of so-called 'women-only' approaches, usually seek to accrue benefits singularly for women without questioning how women had been disadvantaged in the first place (Buchy & Rai, 2008). This approach also plays into the neoliberal argument for encouraging individual success and economic empowerment rather than a focus on multi-scalar and unequal relations of power that Nagar and Raju (2003) refer to.

Babette: What would have been a deal that you would have supported, then, for instance in the space of climate change deliberations? Some feminist scholars are diverting from the usual preoccupation on specific climate change impacts on women (and men).

Kate: Our starting point is climate justice. We differ from those who are focusing on just equal representation but not challenging the patriarchy. We challenge those who cause climate change. We differ from some others in the constituency of course, that's inevitable. We cannot achieve gender justice in an unjust agreement that does not look at common or shared responsibilities for GHG emissions, does not ensure emissions are kept below 1.5°C, or does not make sure the polluter pays. Theses are core principles for us. Nevertheless, we support solidarity amongst the constituency, just like the tradition of Women's Committees in different trade unions. Some may not share the same tactics and principles, but we all at least agree with the need to have women duly represented. However, we're quite clear that there is no empowerment without power. There are no rights without the accountability of those who perpetrate injustice. The UN nowadays uses the language of gender equality and empowerment rather than human rights because you can talk about women's empowerment, which has no legal meaning, and simply seek to increase the number of women in corporations or the number of women engaging in the market. That form of so-called 'empowerment' reinforces, rather than challenges corporate power. The UN has depoliticised the language to placate institutions that have more power than the UN, like the development banks, the IMF, and global corporations.

Apart from their advocacy work with the UN and other global actors, Kate also shares their work with communities affected by widespread palm oil plantations, using the feminist participatory action research (FPAR) approach. She demonstrates how action research can stimulate the growth of local feminist movements.

Kate: We've supported a feminist participatory action research (FPAR) in a palm oil-displaced community in Indonesia. The FPAR explored questions such as: What would the expansion of palm oil mean for women. What impact would it have on their own power? How could they contest the displacement? It will be up to the women to build their own solidarity with the men and others to arrive at common decisions and build their own movement. We've used the FPAR method in climate displaced communities in the Pacific, with

Indigenous women facing landslides and loss of habitat in remote areas of Nepal, with women living in slums and displaced by floods in the Philippines, with women in rural parts of Vietnam, Bangladesh and Cambodia. We support the women to identify their own solutions whilst understanding the global drivers, they build their own movements and conduct advocacy. Their demands all differ and can be very local and practical as well as seeking to collectively pressure governments and corporations globally. But at no time did any of these women say they want to see more women in boardrooms or in delegations.

Another way of co-opting feminism, according to Kate, is the manner with which it has aligned with technocratic practices. Kate refers to this as the practices that use the term 'gender equality' but instead serve to deepen the injustice and to de-politicise. Most state organisations and NGOs with gender equality in their agendas today will have to plan and use log frames to strategise interventions in these displaced communities or that researchers will report the impacts of such displacements, something that Kabeer (1994) might refer to as the 'project trap'. Kate agrees that NGOs face the lure of the project trap but argues that the FPAR method differs in the following way: 'For us it's not about doing a log frame that shows how we will build local wells, it is about whether the local community there would have summoned enough power to speak to the local government and feel that they have authority because they are collectively standing with their piece of evidence from their own research, and have the solidarity of other groups that they've aligned with. It may be that the government will not change its stance, but this is a step forward especially when they experience the capacity to negotiate, be heard and imagine the right to bring about change.'

Log frames and planning have been integral parts of the professionalisation of feminism in many development contexts. There are studies that argue that the very act of professionalisation de-politicises (Nagar & Raju, 2003; Li, 2007; Narayanaswamy, 2016). To illustrate the effects of professionalisation, Morondo-Taramundi (2016) draws attention to the growing gulf between state feminism and feminist movements in Spain. Women's movements have denounced state/institutionalised feminism as complacent and lacking in transformative power, or even colluding with patriarchy. Institutional feminism gradually lost identification with socialist feminist militancy and began to incorporate more ideologically neutral analyses of gender issues according to Morondo-Taramundi.

Kate: Many institutions are desperate to really corrupt the language of feminism. The World Trade Organization, is an example. It has always resisted any idea that it needs to look at human rights and has always explicitly said, 'Our mandate is not human rights.' Last year, it adopted a declaration on gender equality. One that's very clearly just trying to bring women into the system, the system of rules that privilege capital and are designed to ensure the current economic system continues expanding, and they call that gender equality. They are peddling the idea that women's power will come through an expansion of the global economy rather than a redistribution. So absolutely, women should have demanded the right to be part of the economy and still have the right to work and to be paid for work, have the right to property and the same rights as men. But a just economy would ensure the redistribution of paid and unpaid work, not just bring more women into the workforce. A redistribution would allow for greater civic life and sustainable living. We really need to think about how wages have been distributed and what they do to the rest of our lives because sustainable living takes time, and the other aspects of life that are consumption-based take time. The global economic system has valued wages but has not valued the kind of work that is valuable, productive, and care-based that could and should be an essential part of a sustainable economy. Imagine if earlier feminist movements had not just demanded that women have an equal right to paid work, but had instead suggested a redistribution of paid work – a redistribution that should have meant the working week was reduced for all and everyone in the community would have time for social care, for civic life, for the work that is required for sustainable lifestyles.

Becky: So far, we've discussed the effects of professionalising feminism. But where is feminism going at the moment? There seems to be less political unity despite the fact that feminism seems to be gaining high visibility. For instance, I see that prominent women call themselves feminists.

Kate: Yes, there has been a global surge in interest in feminism. But what I guess the effect of this is is to magnify differences between perspectives. If feminist movements were just chugging along in the background, you wouldn't necessarily have contested space so much around what feminism means. But when it's prominent and when invoking feminist narratives can have power, then that power is contested. And so, when very prominent people call themselves feminists, there is more likely to be a reaction to that especially by activists questioning whether those prominent people reflect the interests of all women. Perhaps there's more unity now among the feminist groups that are coming together who try and ensure that feminism is not just for the 1%. Whilst a lot of the very prominent voices are from the 1%, I think there's an increasing response to that and more feminists are saying, 'No, feminism is about shifting the power of the 1% whether they be women or men, to the majority.'

Babette: I've been to a recent conference where women entrepreneurs were talking about women's economic empowerment. These women seemed to come from a different class, they were well-to-do. I did ask them whether they were willing to support the unionisation of their female and male workers. I only received deafening silence and a feverish passing of the microphone to one another.

Kate: Yes, the increasing focus on developing women entrepreneurs – it's quite a ruse. Women have always been entrepreneurs. Women have always traded. So it's not that women lack entrepreneurial skills or ideas, but the global system is making it increasingly impossible for them to survive as small traders.

The majority of world trade is between multinational corporations. And this idea that women are suddenly going to be able to access that world by becoming small entrepreneurs and that that's going to solve the level of inequality of wealth between men and women or between the 1% and the rest is really farcical. There's no evidence of it. There's no evidence that even the smallest level microcredit schemes that are promoted to have women enter into microeconomies have actually changed systems or led to any more than a few individual changes. So those kinds of responses are, again, very neoliberal. They focus on a supposed failure of women – their lack of entrepreneurial skills. All they need is a little bit more skill and then they'll make it. In actuality, the market is rigged against them. It's rigged against women and it's rigged against most men too. The market is increasingly set up for multinationals. And a woman that starts a small microbusiness isn't going to be able to compete in that global system. And in any case, even if they did, it would change the lives of four or five people but not those of other women. I think that those kinds of responses (presented as 'women's economic empowerment') are very much trying to tell us that the system can be just. They're giving us a misleading sense of the (gender equality) problems and how they sit within the global economy. But they're also trying to redirect, I guess, the anger that different groups have around the global economy into making very small band-aid solutions. Instead of being angry that inequalities are rising, that billionaires are making obscene profits from women's labour, that land is being destroyed or monopolised, that austerity measures are privatising public goods, that it is increasingly difficult to imagine a better life for the next generation, women are being encouraged to see their problems as a failure to understand the economic opportunities and encouraged to aspire to increased personal consumption.

From an environmental perspective, trying to bring in more women to be consumers, more women to be billionaires is hardly a solution.

Of course the biggest threat to the current system, the biggest threat to power, is solidarity. An essential part of the neoliberal project is to break solidarity. And that's why, as you say Babette, neoliberal feminists will never support the most proven way for women to increase economic justice – unionising. Even though donors regularly speak about 'results based' development assistance, very few will support unionising even though it has the most proven results for women. Meaningful social progress has only ever been won through solidarity. We can't win real change through professional training and logframes. We can only win change with overwhelming solidarity – that needs feminists and other movements to see that our struggles are bound together. The feminist movement, environmental movements, labour movements, Indigenous movements, all have interests in challenging the existing economic and political system. All risk appropriation by the system.

Conservative, right-wing political parties have been able to position feminism as elitist and exclusionary, a threat to the working class. When centrist parties and global institutions focus purely on increasing women's numbers and profile within the system, they undermine the possibility for solidarity and create the illusion of gender equality.

What would really challenge power and bring about change? It would be all those movements coming together and withdrawing consent from the current system. APWLD has begun seeding the idea of a global strike – a strike designed to reset the system, a strike against neoliberalism, a strike for a new, sustainable, social contract.

Morondo-Taramundi (2016) refers to Antonio Gramsci's notion of the 'passive revolution', where the dominant, often ruling classes, incorporate elements from subaltern groups in order to neutralise them. In juxtaposition, we see an incorporation of feminist discourses (such as 'empowerment' and 'equality') in professionalised and mainstream institutions that Kate alludes to above.

Becky: Do you see the same thing happening with the green economy? There seems to be a lot of optimism over the promises of the green economy for sustainable development.

Kate: When we talk about a green economy, what is it that we're moving towards? It's very easy for corporations to start calling themselves green, and maybe some of them are, but then what kind of justice are they delivering? UN and financial institutions are encouraging what they are calling a just transition to renewable energy, but that just transition too often requires a move away from publicly owned state-based energy to private renewable energy sources. Until this year, Thailand, for example, provided free energy to households of very low energy use where probably you only had a fan and maybe a tv or a very small refrigerator. That is clearly beneficial to women, especially for those coming from the poorest households since they will have to make choices around their expenditure. With a push to liberalise energy and allow new private providers in through renewables, free energy to the lowest consumption households has been lost. These policies are introduced with a green narrative, but actually, a narrative that's quite harmful to both access to public services, to women's rights, and eventually, actually, to the environment, I think, because it takes away the opportunity for publicly funded and managed transition and it also takes away the incentive to use less energy.

Our response to the instrumentalisation of gender within the climate negotiations has been to instead call for a 'Feminist, Fossil Fuel Free, Future. We don't just call for a just transition for workers moving out of fossil fuels, but a just and equitable transition of the entire carbon based economy. That means changing the rules of the global economy – the trade agreements that prevent the regulation of corporations – the tax systems that allow for corporate tax avoidance and undermine public investment – the global supply chain rules that allow corporates to outsource labour and prohibit cross industry and cross country labour organising – the debt rules that give power to creditors to enforce austerity and privatise everything we need to live more equitably and sustainably, and the political influence corporations have over governments and global rules.

Our planet desperately needs this type of feminism.

To conclude

Overall, Kate believes that feminism is increasingly being incorporated and coopted by neoliberalism. She cited examples of the de-politicising effects of the professionalisation of feminism that focuses more on the technical delivery of services rather than political advocacy; of women's economic empowerment as tied to strengthening markets rather than the redistribution of gendered time, wages and assets and the equal exercise of resource rights; of weakening the energy security of poor women as countries transition towards low-carbon energy options driven largely by private sector interests.

Kate's insights chime with recent feminist scholarly debates that view the erasure of feminist politics in the face of neoliberalism. A number of them offer insights worth pondering as they critically engage with growing neoliberalism from different vantage points.

Nancy Fraser (2009), for instance, argues that feminism has itself been complicit in the furtherance of neoliberalism. She cites feminism's earlier critical stance towards the welfare state being paternalistic, which coddled former British Prime Minister Margaret Thatcher's departure from welfare dependence and greater market reliance for the distribution of social services. She also points out that it was feminism's emphasis on identity and culture politics (which she calls the 'politics of recognition' 'the demand to be recognised, to be seen and included, to be given a seat within the system, deflects demands for system change') that eventually deflected attention from redistribution and economic justice. Similarly, Sylvia Walby (2011) argues that feminism must now revert to socialist politics and underscore the claim for redistribution to combat neoliberalism. She emphasises the need to strengthen socialist politics and defend the welfare state to address the adverse and disproportionate effects of neoliberalism on women. For her, neoliberalism 'increases gender inequality directly, through their disproportionate impact on women's jobs and welfare, as well as creating a less hospitable political context for women's effective engagement in the public sphere' (Walby, 2011: 158).

Though not to negate the dire effects of neoliberalism on different groups of women, other scholars are more cautious about the seeming coherence of neoliberalism – or often assuming it as a given – and its almost immense ability to flatten everything on its path. Janet Newman's (2013) research that mapped out feminist activists as they engaged with the state at critical historical junctures in Britain, demonstrates that feminism and neoliberalism comprise dynamic assemblages that constitute techniques and practices that are selectively appropriated as they come into contact with local politics and cultures at particular historical junctures. In short, feminist encounters with neoliberalism are situated (Nagar and Raju,

2003). For instance, Newman cites interviews with feminist activists in the 1970s that critiqued the male breadwinner model in Britain, pushed for child and social care, maternity provision and work–life balance. Newman (2013) notes that this activism can be viewed as first, a selective appropriation of feminist agendas that conferred women with full worker citizen status, but at the same time paved the way for firms to access cheaper, more flexible, less unionised female workforce; second, women are then viewed as integral to advanced neoliberal strategies of sustaining the domestic economy that reproduces conditions of capital accumulation and profit.

Newman also points out that neoliberal projects adapt to feminist activism. As a result of feminist activism in the 1970s, employers have had to bear the cost of equality governance within their firms, parental leave and more complex patterns of work that women demanded as they became full workers. State programs invested in training women enabling them to contribute to the economy and to manage care work. Overall, the gender order remained essentially unchanged. Newman's analysis however points us to the dynamic and complex assemblage of interests and practices, where both feminism and neoliberalism simultaneously gain and lose, drawing attention to where nuanced spaces of political mobilisation might lie.

At this historical juncture of intensive global, national, and local debates over mitigating solutions to climate change and the drastic effects of a worldwide pandemic, it is however fundamentally crucial to keep our gaze fixed on preserving and exercising rights and social justice at the foreground. This means collective organising and strengthening movements to pry open the shackles of corporate capture that often remains unchecked in the search for sustainable solutions.

References

Asia Pacific Forum for Women, Law & Development (APWLD). (n.d.). Our analysis on Patriarchy and G-F-M. *About Us*. https://apwld.org/about-us/our-analysis-on-patriarchy-and-globalisation-fundamentalism-and-militarism/

Bacchi, C., & Eveline, J. (2010). *Mainstreaming Politics: Gendering Practices and Feminist Theory*. University of Adelaide Press.

Buchy, M., & Rai, B. (2008). Do women-only approaches to natural resource management help women? The case of community forestry in Nepal. In B. P. Resurreccion & R. Elmhirst (Eds.), *Gender and Natural Resource Management: Livelihoods, Mobility and Interventions* (pp. 127–150). Routledge.

de Jong, S. (2016). Mainstream(ing) has never run clean, perhaps never can: Gender in the Main/Stream of development. In W. Harcourt (Ed.), *The Palgrave Handbook of Gender and Development: Critical Engagements in Feminist Theory and Practice* (pp. 92–105). Palgrave Macmillan.

Esquivel, V. (2016). Power and the Sustainable Development Goals: A feminist analysis. *Gender & Development*, *24*(1), 9–23. https://doi.org/10.1080/13552074.2016.1147872

Esquivel, V., & Sweetman, C. (2016). Gender and the Sustainable Development Goals. *Gender & Development*, *24*(1), 1–8. https://doi.org/10.1080/13552074.2016.1153318

Fraser, N. (2009). Feminism, capitalism and the cunning of history. *New Left Review*, *56*(2), 97–117.

Fukuda-Parr, S. (2016). From the Millennium Development Goals to the Sustainable Development Goals: Shifts in purpose, concept, and politics of global goal setting for development. *Gender & Development*, *24*(1), 43–52. https://doi.org/10.1080/13552074.2016.1145895

Kabeer, N. (1994). Gender-aware policy and planning: A social-relations perspective. In M. Macdonald (Ed.), *Gender Planning in Development Agencies: Meeting the Challenge* (pp. 80–97). Oxfam.

Koehler, G. (2016). Tapping the Sustainable Development Goals for progressive gender equity and equality policy. *Gender & Development*, *24*(1), 53–68. https://doi.org/10.1080/13552074.2016.1142217

Li, T. M. (2007). *The Will to Improve: Governmentality, Development, and the Practice of Politics*. Duke University Press.

Morondo-Taramundi, D. (2016). Gender machineries vs. Feminist movements? Collective political subjectivity in the time of passive revolution. *Gender and Education*, *28*(3), 372–385. https://doi.org/10.1080/09540253.2016.1169253

Nagar, R., & Raju, S. (2003). Women, NGOs and the contradictions of empowerment and disempowerment: A conversation. *Antipode*, *35*(1), 1–13.

Narayanaswamy, L. (2014). NGOs and feminisms in development: Interrogating the 'southern Women's NGO.' *Geography Compass*, *8*, 576–589. https://doi.org/10.1111/gec3.12150

Narayanaswamy, L. (2016). Whose feminism counts? Gender(ed) knowledge and professionalisation in development. *Third World Quarterly*, *37*(12), 2156–2175. https://doi.org/10.1080/01436597.2016.1173511

Newman, J. (2013). Spaces of power: Feminism, neoliberalism and gendered labor. *Social Politics*, *20*(2), 200–221.

O'Manique, C., & Fourie, P. (2016). Affirming our world: Gender justice, social reproduction, and the Sustainable Development Goals. *Development*, *59*(1–2), 121–126. https://doi.org/10.1057/s41301-017-0066-0

Razavi, S. (2016). The 2030 Agenda: Challenges of implementation to attain gender equality and women's rights. *Gender & Development*, *24*(1), 25–41. https://doi.org/10.1080/13552074.2016.1142229

Walby, S. (2011). *The Future of Feminism*. Polity.

Weber, H. (2017). Politics of 'leaving no one behind': Contesting the 2030 Sustainable Development Goals Agenda. *Globalizations*, *14*(3), 399–414. https://doi.org/10.1080/14747731.2016.1275404

Wilson, F. (1996). Research note. Organizational theory: Blind and deaf to gender? *Organization Studies*, *17*(5), 825–842. https://doi.org/10.1177/017084069601700506

Part 3

The power of gender champions

13 Supporting gender experts

A donor perspective

By Bernadette P. Resurrección and Rebecca Elmhirst
In conversation with Maria Von Berlekom, Eva Johansson, Orawan Raweekoon and AnnaKarin Norling

Both environmental mainstreaming and gender equality sit prominently in development agendas today, almost becoming an established normal in international development after years of keeping it in the margins. In large part, this is being driven by a few development donors. For those of us working on gender and environment for many years, we now seem to be cresting on waves for both intensifying work on gender equality and enhancing efforts for environmental sustainability (including climate) but continue to remain jittery that it may one day ebb due to the volatile political currents. Donor relations significantly influence development and development projects, although in many instances, they are not officially involved in the procedural processes of development planning. Additionally, donor decisions closely align their programming and funding decisions with current international development visions and norms, such as the installation of the SDGs and the UN's Leave No One Behind as the current development compass. For the Swedish International Development Cooperation Agency (Sida), the spirit of their programming is also closely guided by the governing principles of Sweden's first Feminist Foreign Policy (FFP), marking out Sweden as the first feminist government in the world.

Much has been written and discussed about partnerships with the donor community, such as for example the uneasy shifts from participatory, rights-based approaches to more efficiency-driven results-based frameworks in development practice (Eyben, 2014), to donor dependency by women's organisations and questions of autonomy (Petchesky, 2003), and the particularities of recipient behaviour as they chart their way into engagements with donors (Harcourt, 2017). Instead, the interest in this chapter is to briefly explore the evolution of gender equality as a core principle in Sida's partner relations, the unexpected and counter-intuitive experiences of gender mainstreaming of some of Sida's environment and development professionals, and how and whether Sida is conscious of the troubles that gender experts face and experience as they work in technical and environment settings in some partner organisations that Sida supports.

The making of a discourse

The idea of a feminist government was not hatched overnight. It evolved with efforts to arrive at a brand for Sweden in recent years, however gender equality was already publicly represented as a distinguishing trait of the Swedish state as far back as the mid-1990s. Sweden's association with gender equality coincided with its joining the European Union in 1995 that presented Sweden and other Nordic countries as being 'ahead' of Europe in gender equality, receiving an award as the most gender equal state during the 1995 UN women's conference in Beijing, and the intense mobilisation of women to run for office in the 1994 elections after female representation in the government had earlier dropped (Jezierska & Towns, 2018). After the elections, political parties strengthened their gender equality policies and female representation reached an all-time high. This is a case where a growing gender equality rationality is taken up by the state and anchors its identity and the identity of its subjects (Valdivia, 2015). Gender equality was thus high on the agenda, which helped push it into Swedish foreign affairs (Towns, 2002). The world's first FFP was launched in October 2014; it aimed to apply a systematic gender equality perspective throughout Sweden's foreign policy agenda such as in international development cooperation, peace and security, and trade and promotion. 'Brand Sweden' has now been recognised as a utopian dreamscape for progressive liberals (Jezierska & Towns, 2018).

Maria van Berlekom, head of Sida´s Unit for Global Cooperation on Environment and Eva Johanssen, Senior Gender Adviser of Sida further tell us how the principles and practices of gender equality and environmental mainstreaming are deeply entrenched in their operations.

Eva: Since 2014, with our new Feminist Foreign Policy, there is detailed guidance from the Ministry of Foreign Affairs (MFA), to implement and integrate concrete steps for embassies. Sida is a government agency working under the MFA.

Maria: I think that in Sida or in the government as a whole, one doesn't have to argue at all that gender is important. It is so much there. There is also a basic intuitive feeling of what it is all about. Of course, you can argue about the depth of the knowledge and to what extent it is fully applied, but I can't see the need for making the argument for it. The political leadership of Sida is consistent at all levels.

Eva: It's about to ensure that the quality (of gender mainstreaming) is there and so we have updated systems. There is no need to argue the case in any way.

Maria: I was at our embassy in Tanzania and headed our development cooperation operations there. I think looking back, gender equality has been an important goal for Sweden even long before the feminist foreign policy. It is not a new issue. But when the present government was elected in 2015, it was reaffirmed and stepped up. It was an issue with which the government wanted to profile itself. Then they also developed a lot of coherent material. Building on a long tradition and knowledge they stepped it up even more. I would guess even if we had a change of government, I can't see gender dropping out from the agenda, because it's so much part of the Swedish system either way.

Eva: I think that the gender equality principle is very solid. Globally it's received a lot of traction also because having a feminist foreign policy is new. They have been very persistent in promoting it at all levels and in all settings. They also have an action plan every year. That in itself has made a difference. There are benchmarks and they know exactly what they want to do. But also, there has been something of a global vacuum and a backlash on gender equality and human rights, I think, so it was also the right time for Sweden to take a leading role, for example taking the lead in sexual reproductive health and rights. Sweden has taken a very strong position in this area. I think the time was right, it was important that they elevated gender equality.

The unexpected outcomes of gender mainstreaming

Maria points out that Sida's work employs five cross-cutting perspectives that intersect their funding portfolios: (i) poverty alleviation; (ii) human rights; (iii) conflict; (iv) environment and climate change; (v) gender equality. For example, Sida would expect these perspectives to be mainstreamed in huge environmental projects. 'For example, if you take an energy project or an agriculture project, or even an education project, we would both look at how gender and environment and climate change can be mainstreamed in those', she says. She exemplifies the intersecting lines further: 'For example, an education program. We would look at what the challenges and opportunities are for gender equality to be adequately mainstreamed in education. The same would go with environment: what are the environmental risks, vulnerabilities – how can we bring these to the education sector? We would do the same for the energy or disaster risk reduction (DRR) sector. We would have a number of projects, which are sort of 'gender projects' such as supporting UN Women. In that case, we would still look at the environmental risks and opportunities that could and should be mainstreamed in the gender projects. In the case of environmental projects, like the Stockholm Environment Institute (SEI), we would of course look at how they could integrate gender. But in most other programs that are not about the environment, we would talk about both perspectives (and the other three), and in a gender program, we would talk about mainstreaming environment. That's just because we are a development organisation, we are not an environment organisation.'

Sida projects typically require a gender analysis of the target issues for development intervention (e.g., food security, climate change) and gender mainstreaming within its partner organisations. Capacities and interests to achieve these are often uneven. They have put up 'gender helpdesks' to address the need to build capacities through a suite of thorough guidelines, tools and capacity development exercises especially for organisations that have not traditionally addressed gender inequality issues. However, there are a host of unforeseen problems in instituting gender equality in Sida's partner organisations, some of them in women's organisations.

Maria recalls, for example, a situation in Uganda where grassroots women wanted to highlight their need for greater access to natural resources and forest areas, whilst the national women's organisations were not interested in that agenda, because they thought that that had nothing to do with women. She shares her general observations about women's organisations: 'My gut feeling is that "environment" is more of a blind spot among traditional gender activists. More so than the other way around. This is anecdotal, but I know of a Swedish consultancy company wherein there is this false perception that the "proper" gender issues are connected with gender-based violence and conflict as core issues. I sometimes feel that when I talk to gender specialists, there is less "hype" and less understanding around the environment, or it is given less status somehow than what some gender specialists perceive as more "core" gender issues. To me that's a bit problematic. Whilst when I talk with environment organisations that we support, there may not be so much or enough practical knowledge about gender, but there is certainly an awareness and interest in it.'

Babette: I understand where you're coming from, Maria. In April 2018, I attended the Stockholm Gender Equality Forum. Only three out of 28 parallel sessions and eight round tables were dedicated to environment issues.

Maria's reflections above regarding women's organisations and their tensions also remind us of the research of Narayanaswamy (2014) and Nagar & Raju (2003) in India that highlight and explain dissonances between grassroots women's interests and those of women's organisations active in national spaces and arenas. Narayanaswamy (2014), for instance, traces the stream of influence that shape the discourses of national women's organisations, which she argues have been 'colonised' by Global North feminism rather than attending to the more immediate issues of grassroots women around land and livelihoods. She refers to this as a form of elite feminism among national NGOs. In earlier work, Nagar and Raju (2003) echoed the same views, largely questioning whether feminism in the Global North has further entrenched gender inequalities in the Global South. This resonates with postcolonial feminism that questions the universalising tendencies of mainstream feminist ideas and argues that women living in non-Western countries could be misrepresented (McEwan, 2001; Mohanty, 2003). In a way, this could also explain the weak linkages between gender and environment issues within the feminist movement that Maria alludes to above, where feminist groups consider issues of the environment possibly more technical, when compared with gender-based violence (GBV), reproductive and sexual rights issues, and gender wage gaps, which have been the conventional core targets of feminist engagement. However, linking gender inequality and climate change in policy action has increased in many women's organisations and environmental programs recently and often occupy a niche within their operations. For instance, the recent launch of an improved Gender Action Plan by the United Nations Framework Convention on Climate Change (UNFCCC) in the Conference of Parties (COP) 25 in Madrid was celebratory, 24 years since the 'Solidarity in the Greenhouse' women's meeting in COP1 in Berlin in 1995, whose momentum could not be sustained (Resurrección, 2013). Another example is the Sida-supported EmPower program within the regional office of UN Women in Asia-Pacific in partnership with the UN Environment Programme (UNEP) that tackle issues on climate change and renewable energy in Asia and the Pacific. The Asia Pacific Forum on Women, Law and Development (APWLD) has created a program on feminist participatory action research (FPAR) that seeks to understand gender dimensions of climate change in communities and advocates for a Feminist Fossil-Free Future (see Chapter 12 with Kate Lappin). The recent Asia-Pacific Declaration on Advancing Gender Equality and Women's Empowerment following the Beijing+25 Review has prioritised environmental concerns. Whether and how these niches of gender and environment programs and platforms link with other parts of their larger organisations and conversely contribute to integrating environment in future gender agendas certainly bears watching. It will probably take some time before these silos are dismantled.

We were also curious whether Sida is conscious of the deep-seated power relations in their partner environment and development organisations, and how gender professionals are entangled in these relations. Many times, these are the very barriers that constrain meaningful change within these organisations, especially since gender inequality is often recognised as a problem external to the focus areas and concerns of these organisations themselves. The next section discusses this at length.

Making space for gender experts in Sida's partner organisations

A good number of Sida's partner organisations have employed gender experts to do gender analysis and gender programming, a few of whom are also contributors to this book. Becky asks: 'What happens if you meet with resistance to gender equality as a value among your partners?'

Maria: That's a tough question. But sometimes our advantage is that we have on-the-ground knowledge about what's happening. This enables us to make inputs and frame the discussion in a way that can make gender make more sense – more contextual and more understandable – for our partners' use. One would have to re-phrase certain wordings when working with certain organisations and actors just to get them through and understood. For instance, we are gently pushing on the issue of gender but still having respect for SEI's core mandate. I think we would take the same approach for other organisations. Sometimes we should perhaps be a bit more pushy, and sometimes we have to be less pushy. It's difficult to say. But I think that the basic thing is that we try to show the way, provide dialogues and inputs, but trying to maintain an encouraging approach rather than a demanding approach.

Eva: In some countries, I worked as the gender adviser of UNFPA who Sida supports. If you're working on a very sensitive area like sexual reproductive health and rights, you have to explain what difference this makes. It has to be rights-based. In some contexts, it gives added leverage if you can have a narrative which is broader than health and rights issues. If you can explain that it has economic benefits, for example. Child marriage has negative income impact on the family and the country. It will be a short-term gain if you marry off the girl, but it will be a long-term gain if you put her in education. So, it's a little bit about the kind of narrative that's very important. This can neutralise challenging questions, make them more acceptable in a way, and just needing to put these issues in an (acceptable) context.

Sida's Asia-Pacific regional office supports about 45 partner organisations, and about half of them are technically oriented environmental organisations. According to Orawan Raweekoon, the regional office's Human Rights Officer, 'Our priority now is to work with environmental organisations to get gender within the different programs they have.' AnnaKarin Norling, then Research Advisor in the regional office at the time of this interview, tells us that the reception of gender equality concerns has been different for different organisations in the region. She says, 'There were times that I saw some resistance to incorporate gender from these organisations. And their first reaction to begin gender mainstreaming is usually to count women or bringing in women researchers. I constantly say that it's not only about counting women, but also about including a gender perspective to the research.' For instance, she recalls: 'I remember one researcher from one of our partner organisations who said that gender equality is not a problem in the Mekong region – it's economic development and environment. He drew this insight directly from shared views within his organisation.' Other organisations, meanwhile, welcomed the added push from Sida since they had been minimally doing gender-related work already in the past. 'We provide the additional push for some organisations whose members express keen interest to apply gender equality strategies or have been doing this work but not in a major way', AnnaKarin points out.

Babette: Could you provide us more examples of how you engage with partner organisations in the field of environment and development, in embedding gender in their work? How did they react to the need to bring in gender?

AnnaKarin: I recall that a researcher gave a presentation on his organisation's analysis of the situation where a geothermal plant had been developed in a district in the Philippines and the effects of this on nearby communities. He said that the overall rate of elementary level enrolments and graduation had increased in one of the district schools. So, more kids have gone to and graduated elementary school. Then I asked him whether this was true for both girls and boys. And he gave me a perplexed look and said he didn't know. And I said this is why I wanted to have a gender perspective in your research. And he said: 'I have to find this out and go check.' And then when we talked later over dinner, he said that he was really excited about finding out more and understood why he hadn't, or why he was blind to it. I then saw how it clicked [laughter]. So, it's not always obvious, but people can learn to be more conscious.

> I realise it may be difficult for some organisations to adopt a gender strategy. But we at Sida are very serious about this being *our* strategy. These are the results that we are contributing to. And if you don't do this you can't have funding from us. But at the same time of course I hope even if it is donor-driven that along the way they will understand. And that's what our dialogue with partners is all about.

AnnaKarin's example on the effects of the geothermal plant reflects many things: the virtues of dialogue between donors and their partners that may lead to 'epiphany' moments; gender blindness among technical specialists that usually sidestep the social outcomes of technical interventions like the installation of a geothermal plant which was intended to be a climate change-mitigating measure. The question that is hardly asked in these technical and infrastructure-centred contexts is, 'What's in it for women and men, girls and boys?'. The gender perspective that AnnaKarin points out to the researcher challenges the assumption that technical solutions will inevitably have trickle-down benefits for all. This perspective inevitably draws attention to how technological fixes may exacerbate existing relations of inequality and power.

Rosalind Eyben (2014) underscores the need to be power-sensitive as she reflected and described her long journey as a former member of the UK's Department for International Development (DFID), another donor organisation. She argues that current professionalised results-based methods and the emphasis on best practices to report successes and 'quick wins' of funded projects often lose sight of power and politics. She recalls a story where two separate visits with Ethiopian farmers were made by a former development minister and herself, and which yielded different insights and viewpoints. Her farmer hosts who grew cotton as a main cash crop lamented that they could not get good prices for their crop and were already in deep debt for farm inputs, so they were compelled to stop growing it.

Rosalind explained to her hosts at the time that, in the US, farm subsidies were reducing world cotton prices, which could explain why cotton sales in Ethiopia dropped. She doubted very much whether the same insight and analysis emerged during the minister's visit. Instead, the civil servant accompanying the minister reported that aid has become crucially important to deliver services especially in view of the farmers' poverty. 'Nothing was said about power in the global political economy that may have negatively affected their hosts' chances of earning a living. A "best practice" approach maintains the invisibility of such issues by rendering poverty reduction a technical activity; the intervening aid agency absents itself from any analysis of the power relations that keep some people in the world poor and others rich. Yet by seeking to do something about extreme poverty though the delivery of tangible results, and without addressing the relations of power – within communities, nations and globally – of which people's poverty is a consequence, international aid agencies may find it difficult to achieve their own goals' (Eyben, 2014: 149–150). Similarly, in climate change and wider environmental debates, there is a growing concern that priorities for planning, action and funding lie in advancing infrastructural and technical solutions (Nightingale et al., 2019; Eriksen et al., 2015; Ensor et al., 2019).

Orawan tells us about another partner organisation, mostly engaged with sustainable development activities and programs in the fisheries sector, where a process to include gender issues in its vision and mandate has been gradual and has taken very long. It is an organisation that has been in existence for more than 50 years, and its chief concern is fishery management, both marine and aquaculture. Through this example, she describes to us the challenges of introducing gender issues in technical organisations.

Orawan: We tried to influence the organisation from the very beginning of our funding support, so we agree to say that this may be a donor-driven process and they have to include gender equality in the proposal. It normally turns out to be a checklist culture written in the document. But when we ask them in detail what they mean by 'gender equality', they are not able to explain what they plan to do. This is the challenge we have from the beginning and for one fisheries organisation, it took them almost five years to come up with only one gender strategy of their organisation. It's taken us a long time in dialogue. But the funding I think sets the condition for them to work on gender. That's the key.

AnnaKarin: We are very open that this is our strategy. These are the results that we are contributing to. But at the same time of course I hope even if it is donor-driven that along the way they will understand (gender issues). As Orawan points out, we go far to support and build capacities and also to dialogue to make change happen.

Becky: What has facilitated deeper engagement? Is it just the money or is there something else you have to have to do to get that deeper engagement?

Orawan: If I may chime in, I think it's not only the money that motivates, but we also provide capacity building. We have held one workshop with the leaders of our partner organisations. The workshop provides awareness-raising on the basic principles of gender equality. This is because most of the challenges arise from no buy-in from the top level of the organisation. This is another strategy, so it is not just only money that motivates. We also have a regional gender help desk that provides hands-on support to those who would need the capacity building. For example, this help desk can help train on gender analysis or formulating monitoring and evaluation (M & E) indicators for the organisation. So that's another service or form of support that the regional office provides. We hold workshops and train top management and raise their consciousness on the importance of gender in their organisations and institutional programs. These workshops where the heads and directors of our partner organisations are invited to ensure that the mandate for gender mainstreaming comes from the top. The attitudes and mindsets of the leadership and top management level need to change. There is need to bring gender equality to the top agenda of the organisation. We also try to make them understand the importance of allocating a budget for gender-related programs and activities. Otherwise, the gender specialist will have to fight for resources and it's going to be tough for them to be always struggling.

Orawan also tells us that five-year program agreements with Sida require that the partner organisation employ a gender expert. Once employed, it's probable that gender experts experience multiple dilemmas, chief of which is a sense of isolation especially in highly technical settings. As contributors in this book say, their knowledge is often tacitly considered supplementary to the authority of the technical sciences, and this is exacerbated by the fact that transforming gender equality is not the core business. It is often challenging to incorporate a gender perspective in an organisation that does not prioritise changing unequal power and gender relations. Do donors like Sida realise and understand the struggles these experts experience? Additional questions may arise, given the goal of environmental sustainability of these organisations: How can gender considerations add value to the mission of conserving and protecting the environment or resources? In short, such a concern is more directed at identifying the *usefulness* of including gender in improving biophysical and environmental outcomes, but less on ensuring equal benefits for all, including for women as a group. This also focuses on instrumental gains, or in short, a focus more on what women can do for environment and development rather than what development can do for women (Cornwall, 2016; Ferguson, 2015).

In other chapters of this book, contributors express their struggles in science-led organisations because at core, gender equality is a political goal that seeks to transform unequal relations of power to benefit those marginalised, realise equitable access and control of resources and equal rights to decision-making and overall well-being (Leach, 2015; Rocheleau et al., 1996). Their work is not intended to serve the ends of environmental protection, nor to boost the business profits of firms, although there are efforts to bend it towards these ends. This could be where tensions and contradictions possibly lie.

Concluding remarks

The chapter also showed us some of the struggles of gender experts from the perspective of donors. How can they further support gender experts who may be dealing with knowledge hierarchies and isolation in technical work settings? How can the experiences of gender experts provide useful lessons for donors in moving the gender agenda forward in these rather difficult settings?

The examples in this chapter provide us with important food for thought about how members of a donor organisation perceive and enact their gender equality mandate and work with partners. In relation to this mandate and drawing recollections from her years as a DFID employee, Eyben (2014) underscores the importance of recognising the positionality of donors in a landscape of power relations. For her, recognising forms of visible (or formal) and informal (or invisible) power is crucial: visible power is seen through formal institutional arrangements for policy making such as increasing the numbers of women in parliament, whereas invisible power is able to shape and influence policy agendas from a position at the 'rear,' and often comes from internalised norms and beliefs 'about how the world should be' (p. 130). Donors are often influential through invisible power, according to Eyben. What also emerges from these conversations is that donors are profoundly influential and can potentially be real agents of change with respect to enacting a gender equality agenda in environment-oriented organisations. They can also conversely bring environmental issues to the attention of feminist organisations. They can potentially support gender experts navigate the difficult technical terrains that they are asked to influence and transform. However, in studying women working in NGOs across the globe, de Jong (2017) discussed donor relations and alluded that donors unconsciously position themselves as external actors standing 'outside' the development realities that their organisations are trying to transform through development aid.

Eyben (2014) suggests that it may help to be co-reflexive: to see ourselves through the other person's eyes and put ourselves in others' shoes. Additionally, de Jong (2017: 93) suggests the need to critically examine the investments we make on ourselves that maintain an external presence from those we partner and the unequal power realities they face. A meaningful connection can lead towards a more transparent and honest appraisal of the steep problems that gender experts and supportive donors face in environment and development practice, and hopefully fuel and affirm greater solidarity in collectively imagining an equal and more sustainable world.

References

Cornwall, A. (2016). Women's empowerment: What works. *Journal of International Development*, *28*(3), 342–359.

De Jong, S. (2017). *Complicit Sisters: Gender and Women's Issues Across North-South Divides*. Oxford University Press.

Ensor, J. E., Wennström, P., Bhatterai, A., Nightingale, A. J., Eriksen, S., & Sillmann, J. (2019). Asking the right questions in adaptation research and practice: Seeing beyond climate impacts in rural Nepal. *Environmental Science & Policy*, *94*, 227–236. https://doi.org/10.1016/j.envsci.2019.01.013

Eriksen, S. H., Nightingale, A. J., & Eakin, H. (2015). Reframing adaptation: The political nature of climate change adaptation. *Global Environmental Change*, *35*, 523–533. https://doi.org/10.1016/j.gloenvcha.2015.09.014

Eyben, R. (2014). *International Aid and the Making of a Better World: Reflexive Practice*. Routledge/Taylor & Francis Group.

Ferguson, L. (2015). 'This is our gender person.' The messy business of working as a gender expert in international development. *International Feminist Journal of Politics*, *17*(3), 380–397. https://doi.org/10.1080/14616742.2014.918787

Harcourt, W. (2017). The making and unmaking of development: Using Post-Development as a tool in teaching development studies. *Third World Quarterly*, *38*(12), 2703–2718. https://doi.org/10.1080/01436597.2017.1315300

Jezierska, K., & Towns, A. (2018). Taming feminism? The place of gender equality in the 'Progressive Sweden' brand. *Place Branding and Public Diplomacy*, *14*(1), 55–63. https://doi.org/10.1057/s41254-017-0091-5

Leach, M. (2015). *Gender Equality and Sustainable Development*. Routledge.

McEwan, C. (2001). Postcolonialism, feminism and development: Intersections and dilemmas. *Progress in Development Studies*, *1*(2), 93–111. https://doi.org/10.1177/146499340100100201

Mohanty, C. T. (2003). *Feminism without Borders: Decolonizing Theory, Practising Solidarity*. Duke University Press.

Nagar, R., & Raju, S. (2003). Women, NGOs and the contradictions of empowerment and disempowerment: A conversation. *Antipode*, *35*(1), 1–13.

Narayanaswamy, L. (2014). NGOs and feminisms in development: Interrogating the 'Southern Women's NGO.' *Geography Compass*, *8*, 576–589. https://doi.org/10.1111/gec3.12150

Nightingale, A. J., Eriksen, S., Taylor, M., Forsyth, T., Pelling, M., Newsham, A., Boyd, E., Brown, K., Harvey, B., Jones, L., Bezner Kerr, R., Mehta, L., Naess, L. O., Ockwell, D., Scoones, I., Tanner, T., & Whitfield, S. (2019). Beyond technical fixes: Climate solutions and the great derangement. *Climate and Development*, 1–10. https://doi.org/10.1080/17565529.2019.1624495

Petchesky, R. P. (2003). *Global Prescriptions: Gendering Health and Human Rights*. Zed Books.

Resurrección, B. P. (2013). Persistent women and environment linkages in climate change and sustainable development agendas. *Women's Studies International Forum*, *40*, 33–43. https://doi.org/10.1016/j.wsif.2013.03.011

Rocheleau, D. E., Thomas-Slayter, B. P., & Wangari, E. (1996). *Feminist Political Ecology: Global Issues and Local Experiences*. Routledge.

Towns, A. (2002). Paradoxes of (in) equality: Something is rotten in the gender equal state of Sweden. *Cooperation and Conflict*. 37, 157. https://doi.org/10.1177/0010836702037002975

Valdivia, G. (2015). Eco-governmentality. In T. A. Perreault, G. Bridge, & J. McCarthy (Eds.), *Routledge Handbook of Political Ecology* (pp. 467–480). London and New York: Routledge.

14 Gender equality work at USAID

Mandatory as applicable

*By Kai Spratt and Charles 'Will' Lewis II**

The wider context of development aid

No chapter on the role of donor involvement in advancing gender equality in environment and development would be complete without at least a cursory wade into the current global landscape of development aid and cooperation. Development aid involves both complex financial transactions and political processes of defining target agendas and priorities for funding. The consensus among a number of scholars is that the donor community narrative has significantly changed from poverty alleviation to 'shared prosperity' by means of active involvement of the private sector as a partner to stimulate economic growth (Campbell & Teghtsoonian, 2010; Eyben, 2015; Mawdsley et al., 2018; Murray & Overton, 2016; Roberts, 2014). Economic growth is therefore the vehicle by which prosperity and development is being envisaged.

Following the collapse of global capital markets in the 2000s, the principal means by which such markets were reconstructed was through the making of large financial stimulus packages, some of which fell under the umbrella of overseas development assistance (ODA). Donor states began to offer safety nets to private corporations to put economies back on track. Examples of these packages were the bail outs to large motor corporations in the US and banks in the United Kingdom, which was reminiscent of the post-war era, and thus thought to be 'retroliberal' (Murray & Overton, 2016).

Specific features define the current retroliberal nature of development aid, which Murray and Overton (2016) and Mawdsley et al. (2018) describe as follows: (i) economic growth is the principal target of aid causing earlier poverty reduction goals to recede; (ii) increasing dominance of donor state self-interest compared with previous periods when national interest was separate from principles of aid donation (or the 're-entangling of diplomacy and aid'); (iii) emphasis

* Based at the Regional Development Mission for Asia, USAID, in Bangkok, Thailand, Dr Kai Spratt, Senior Regional Gender Advisor, and Charles 'Will' Lewis II, Mission Gender and Inclusive Development Advisor, lead efforts to integrate gender equality, female empowerment and social inclusion into the programming, processes and implementation of US foreign assistance in Asia. The authors' viewpoints expressed in this chapter do not necessarily reflect the views of the United States Agency for International Development (USAID) or the US government.

on infrastructural projects that are intended to enhance and boost outward trade; (iv) the use of aid in ways that align with security and military goals. These observations are buttressed by a recent Overseas Development Institute (ODI) paper on donor motivations measured through a Principled Aid (PA) Index that ranks 29 ODA countries according to whether their aid support is 'principled,' or supporting 'long-term national interest in a prosperous, stable and secure world ... without sacrificing global development ambitions' (Gulrajani & Calleja, 2019: 7). In short, the PA Index was intended to show whether donor states were straying from addressing development goals as they were conventionally framed. It is not coincidental that the making of the PA Index comes at a time when more donors are releasing sub-optimal aid allocations for reasons other than development, detracting resources and efforts away from the core and primary objectives of global sustainable development (Girod, 2008).

In the particular case of the US Agency for International Development (USAID), Roberts' (2014) study highlights the financial aid transactions and the rise of contracting US firms as the most significant relationship constituting US development aid today, thus giving impetus to a growing 'industrial-development complex'. USAID has been observed to gradually morph into a sub-contractor of the substantive side of its development mission to contractors and hollowing out the technical capacities of the agency itself retaining mostly administration and management personnel. 'USAID has been turned into a kind of distribution channel where the expertise that had taken years to build up has now migrated outside of the government' (Roberts, 2014: 1035). Roberts' paper also points to a regional geography of big and affluent players who receive USAID grants and contracts and are located within commuting range of Washington DC known as the 'Beltway region.' For example, 56 per cent of development aid grants to Haiti went to organisations in the Washington DC region (Johnston & Main, 2013).

From records pieced together by Roberts (2014), USAID contracting is dominated by for-profit companies, some of which have grown to be large corporations. Rajiv Shah, President Barack Obama's choice for USAID's administrator himself said in a 2011 speech that signalled his vision for internal reforms, indicated the worrying trend of contracting big firms, said: 'USAID is no longer satisfied with writing big checks to big contractors and calling it development. There is always another high-priced consultant that must take another flight to attend another conference or lead another training.' Shah also argued that development firms were more interested in keeping themselves in business than seeing countries graduate from the need for aid (Norris, 2012). To this, adds Roberts (pp. 1045, 1046):

> It is particularly poignant that dollars allocated to make a positive impact in the lives of the world's poorest often never make it beyond the Beltway region. These projects' many reports, and the water pipes, health clinics, and studies they assemble, can be taken to indicate self-evident 'spaces of development assistance.' But, just as surely, the office parks, cocktail parties, awards luncheons, networking seminars, and multi-million-dollar homes of

> development contractor CEOs in Northern Virginia are themselves 'spaces of development' or more properly spaces of development capital.

All these trends in development aid have implications on contemporary efforts to advance gender equality within USAID. It must be recalled that USAID was an early pioneer on women, gender and development programming in the 1980s and a precursor to the gender mainstreaming imperatives expressed and adopted by the Beijing Platform of Action in 1995 (Rathgeber, 1995; Overholt et al., 1985). Gender equality remains a policy within USAID, but current resources channelled to achieve this do not live up to the long history of its policy ambitions. As Campbell and Teghtsoonian (2010) point out, development aid is a set of ruling relations, thus current discourses on gender and development that focus on economic growth can define agendas and operations, posing risks that gender equality efforts may altogether lose their transformative edge. In tracing the trajectories of gender and development over decades, Eyben (2015), singles out the World Bank discourse on 'gender equality is smart economics' as aligning with the current centrality of economic growth in development aid, and thus the increasing propensity of primarily framing women as economic agents without efforts to address wider social and political economic structures that drive their inequality in the first place. Gender equality itself has been made 'a business case,' whereas feminists have long argued that economic growth as a goal and as a reality has not turned around gender and other forms of inequalities (Cullen & Murphy, 2018; Ferguson, 2015; Kabeer, 2013; Fraser, 2009). This poses serious dilemmas for gender experts who have to navigate 'the business case' ('gender should generate bucks' approach) to legitimate a gender equality agenda.

Like other development organisations these days, USAID as an early and celebrated pioneer of gender and development is perhaps now but a shadow of its former years: gender remains mandatory, but at the same time, expendable. There also seems to be little effort to turn around conventional practices of hiring personnel with technical portfolios, many of whom keep to their areas of expertise resisting any inroads of gender integration similar to many other technical organisations. Kai Spratt and Will Lewis, gender advisors, reflect on their day to day work in the Asia regional office and with the country missions of USAID. They highlight spaces for both contestation and negotiation; they show us how critical reflection as gender experts can indeed be an opportunity to see the political.

The historical evolution of gender work at USAID

The core business of USAID is to 'promote and demonstrate democratic values abroad, and advance a free, peaceful, and prosperous world. In support of America's foreign policy, USAID leads the US Government's international development and disaster assistance through partnerships and investments that save lives, reduce poverty, strengthen democratic governance, and help people emerge from humanitarian crises and progress beyond assistance.' The Agency was founded by an act of Congress in 1961. The Agency is headquartered in

Washington, DC. Country-level presence is in a 'mission' usually co-located with the US Embassy. USAID works in South and Southeast Asia, Africa, the Middle East, Central Asia, Eastern Europe, and Latin America. USAID works in multiple sectors in every country: democracy, human rights and governance (or DRG); economic growth and trade; health; education; natural resources management and; energy and infrastructure. In its beginning, the business of development was considered mostly technocratic problem-solving, and practitioners just assumed they were reaching both men and women. It was soon apparent that ignoring gender was not a successful strategy.

Efforts to address gender inequality have been going on for decades at USAID. In 1973 the US Congress passed the 'Percy Amendment' to the Foreign Assistance Act, requiring focused integration of women into national economies through US bilateral assistance. USAID responded the following year by setting up the Women in Development (WID) Office, which championed gender-specific analysis, the usage of sex-disaggregated data, and strategies for explicitly benefitting women and girls. This era was very academically focused and research heavy. The WID emphasis would last through the mid-1980s, but the 'Percy Amendment' is still in effect today.

Then came the Gender in Development (GID) approach in the 1990s. These efforts were housed in a very small office with only a few staff whose job was to advocate for gender equality, train staff and do analyses and reports. The common perspective during the period was that 'gender equals women.' The focus was on the development of tools and processes, but much of the research focus of the WID era went away.

In 2012, with the overall reinvigoration of the policy office at USAID, we got the Gender Equality and Female Empowerment Policy (known as the Policy) that required us to integrate attention to gender into the work that we do in a much more formalized way. In the same year, we also got Agency guidance on how to implement the Policy (procedures called the Automated Directive System Chapter 205 (ADS 205) that strategically laid out a set of requirements for projects: gender analysis, use of gender sensitive indicators; requiring that missions have gender advisors or points of contact; that findings from gender analysis should inform project design and be integrated into contracts. We don't use the term 'gender-mainstreaming' at USAID because we learned from the previous generations of working on gender in the development field that this often meant throwing in the word 'women' into a document a few times and you've gender-mainstreamed. The Policy and ADS 205 require that we be very specific about the gender inequalities we want to address in a project (we use the term 'gender gaps'), so activities can be designed to address those gaps and measure progress towards closing the gaps.

USAID does development largely by disbursing grants and awarding contracts. In many cases, it releases calls for proposals, and organisations bid to win contracts to undertake the specified work (Roberts, 2014). The funding cycle for USAID grants and awards follow a 3–5 year cycle and is a result of the way we are funded by the US Congress. The idea of long-term investments projects is usually not entertained. In general, USAID/Washington is the point of contact for Congress, for communication in Washington and the homeland on what USAID is doing around the world. In the field we may get urgent request for an answer to a question about what is going on in country X so that can be used for testimony to a meeting with a congressperson or committee who has specific questions about our work in a specific country or on a specific topic, like disabilities or trafficking or natural resource management. Our Administrator will meet with Congress, but so too will many others in the leadership hierarchy. Our job at the Mission is to provide them with updated information for these meetings. On the ground our senior American and local staff work to build bridges and relationships with the national and other governments through sponsoring technical workshops, holding consultations with governments and local organisations on developing of new five-year strategies.

At the Mission-level, the various Washington-based initiatives and directives are evidenced in each Country Development Cooperation Strategy, which describes five years of strategic engagement with host countries to resolve complex development challenges. It is mandatory within USAID that a gender analysis is done to identify regional and country-wide trends and gender equality gaps that the mission will address over the life of the strategy. It is expected that the gender advisors have a key role in conducting the analyses and in identifying how best to integrate the identified gender gaps, although the ultimate strategic decisions should be made by the technical office staff.

The work of gender advisors in the regional and bilateral missions

In the last 15 years, USAID has employed a more diverse set of people with degrees in development policy or business, agronomists, even public relations specialists. But 90 per cent of them are not social scientists.

Our job as gender advisors is to bring the process as described in the Policy and ADS 205 to the people that we work with to design and implement new projects. We can't do it all. We have to support our colleagues who are technical experts, program specialists, monitoring and evaluation advisors, administration staff, to understand how to do gender analysis and integrate activities to reduce gender gaps into our development work. We do not work with feminists every day. We work with people who truly care about gender equality, with people who are sympathetic, people who are hostile to the idea of gender equality and some people who still need to be 'sold' on the idea that it will make a difference to them achieving their project's goals. We are still trying to convince most people that our development work should address inequalities based on gender and other kinds of intersectional inequalities. Our colleagues will ask, 'What does social inclusion have to do with my energy project?' Or people will say, 'Well, this is an environment project, it's not a gender project. We can't take money from our budget to address that.' Our policy says regardless of the how projects are funded there has to be activities in that project that address gender disparities, gender-based violence, and women and girls' empowerment. Colleagues are under considerable time constraints to design a project and release the proposal for bidding so there is often reluctance to spend the time and money to do a rigorous gender analysis. And so, we work under very significant time and resource pressures with few human resources with appropriate expertise.

Every regional and bilateral mission has (or should have) a gender advisor, according to ADS 205. This person can be hired or assigned if already working in the mission. The Regional Development Mission for Asia (RDMA) USAID, or the regional mission in Asia, has a very empowered Foreign Service Officer (Will) who is the gender advisor, though only 20 per cent of his time is dedicated officially to that role. He is privileged in that he can take on his American colleagues as an equal. The senior regional gender advisor (Kai) supports and mentors the bilateral mission gender advisors across Asia and, as an American with 20 years of experience and a PhD, is accorded time with senior leadership in every mission that the bilateral mission gender advisors are not privileged to have.

Becoming a gender advisor was rather accidental in Will's case. While he was a sociologist who specialised in gender and sexuality, with a sub-focus on international conflict and globalisation, as a US Foreign Service Officer (FSO), he was a generalist working on project design, research and monitoring and evaluation. However, as he worked on designs, it became rather obvious that he needed to develop systems, tools and templates to improve gender and social inclusion (GESI). His supervisors were gung-ho to allow someone with expertise to work on integrating GESI across development interventions, as long as he remembered that his 'real' job was in program and project design. He carved out a niche for himself, and, after eight years, he is still one of the few FSOs (and only male FSO) working on developing systems for integrating gender equality and female empowerment across development interventions. There is not a career track for a GESI Advisor in his position per se, so his work on gender equality is still rather at the mercy of future supervisors. As such, his peers are generally local staff members from other offices in Asia and the experts based in Washington, DC – they meet infrequently.

It should be noted that being a male Foreign Service Officer working on GESI integration and the power inherent within the institutional position allows him to shape the system in ways that might be more difficult for local staff. To be honest, he comes from a position of privilege. In Asia, 95 per cent of gender advisors are local staff or Foreign Service nationals (FSNs) stationed in the bilateral or country missions.

Most FSNs, like Will, do not have 100 per cent of their time focused on being the gender advisor: the average in Asia is 30 per cent. While there is no official or standard job description for a gender advisor, their job is to guide mission colleagues through every phase of gender integration from doing gender analysis, project design, to reviewing proposals and monitoring plans, assisting with development of gender sensitive indicators, staff training, providing feedback on outcome evaluations to reporting on mission progress to Washington. The majority of mission gender advisors have had no training on gender before being 'voluntold' to be the gender advisor and little if any training or mentoring after being assigned to that role. They may have been chosen because they are the women in the mission, they were available, and they may have had an interest in gender equality. They are not monitoring and evaluation experts. They may or may not be comfortable being trainers. They are generally not subject matter or technical experts but are often program management staff. They generally have a mid-level professional ranking. They have a lot of responsibility but no authority. This makes their job very difficult. Their advice may not be heeded by technical experts who see them as 'support staff' and not technical advisors; they cannot compel higher ranking staff to conduct mandatory gender analysis or insist that a project report on gender sensitive indicators. They have to navigate the intersectional hierarchies within their own mission. That they persist is remarkable. Kai and Will have advocated for at least one Asia-wide gender advisors' training yearly for gender advisors in the bilateral mission offices. Think of the impact gender advisors could have if they received comprehensive training and mentoring and were given authority to hold people accountable. Our work would look so different and we would be making a real difference. That they haven't had comprehensive training and mentoring speaks to the importance of gender integration to the Agency.

Many of the FSN gender advisors are marginalised within their own offices, and their success in the role comes from not rocking the boat too much, especially in negotiation with American supervisors and other Foreign Service Officers. They are often limited to simply advising or trying simply to convince. One of the local gender advisors on seeking some advice for how to convince colleagues to do a mandatory gender analysis said, 'I myself have little voice or even never dare to push for that. [We are just pushed] to get the [project] contract out.' Cultural norms regarding gender and class, debate, interactions with higher ranking staff, giving feedback, and workplace behaviours shape their ability to be as effective. All this negatively impacts the work because it structurally silences our local staff who are the best resources for understanding development issues and the potential consequences for development solutions.

Thus, colleagues' reaction to the mandate to integrate gender is very mixed and varies from mission to mission. You may go from one office which has done a good job to another mission that hasn't. And so, it's an uninformed approach. We have a sense that all depends on the mission. There are many people who really want to do well. They often say to us that we would like to learn more. We would like to be advocates. But they just don't have the resources. And they don't have the research to actually understand gender gaps in the work that they do. We in the regional mission are trying to think through how to make the process part easier so that they can do true analysis and integration.

Being a gender professional implies expertise, experience, and consistency in purpose and professional development. In many organisations, gender equality work is often viewed as part-time work and assigned to someone with little formal training who has 'spare' time. In development work though, we are striving to understand amazingly complex social issues with the ultimate goal of working with partner countries to shift behaviours. Mistakes and a failure to properly understand cultures and people's needs could result in harm. As such, it demands a level of time, resources, expertise and accountability beyond a part-time approach or general advocacy- in short, a professionalism borne of competency rather than availability.

The work we perform may be immensely frustrating, and we aren't certain how much gender advisors in the missions have a sense of hope that we can and will and are making a difference. For many it is such a small part of the work that they are tasked to do they may not get a sense of what their efforts are accomplishing. Some aren't feminists, so the urgency of striving for gender equality is not a priority – it's just part of the job. For the ones who are feminists we have made an intellectual bargain, as one of the gender advisors aptly said, 'If I keep at it, even though I cannot speak my mind, I can still make a difference.' Maybe this is the very definition of 'hope' under constrained circumstances.

Accountability for gender equality

A 'gender professional' has to be a feminist – living in a world where there is equality for all has to be a deeply held personal commitment. Kai became a feminist after reading Betty Friedan's 'Feminist Mystique'; it absolutely changed her teenage ideas of what her life could be (Friedan, 1963). From then on she looked at situations through Diane Elson's lens 'Who's here? Who's not here? Why?' (e.g. Elson, 1995). What excited Kai about being a gender advisor at an international donor agency was a giddy optimism that if we really did ensure all our projects were addressing gender inequality, if we were really were working with local actors to address gender based violence and empower women and girls we would be meaningfully contributing to the transformation of social inequalities in the communities where we work. That's what our gender equality and female empowerment policy laid out. That is why Kai joined. That is not, it turns out, how it works. There are many really dedicated people at our Agency – some of whom have been at it for decades – working towards our policy goals but until there are real human and financial resources dedicated to transforming the way we address inequality our work is mostly symbolic. Our policy has never been costed out to understand what it will really take in human resources or staff capacity building to make it a reality. Somehow, through passive osmosis, all staff are supposed to understand how and what to do to integrate gender into their work.

Being a gender advisor and a feminist, one should also not fear to turn this critical perspective on oneself and one's organisation. While sexual harassment and overt discrimination are handled openly and acknowledged, more subtle bias is difficult to address institutionally. For example, there is the belief that social inclusion is a matter for women and minorities and, thus, marginal to overall development efforts. Thus, if you add a few women to one's project or address just opportunity, the action is sufficient. Deeper inequalities remain unchallenged. Mandatory training and advocacy help staff members understand why addressing inequality is important for achieving good development outcomes, but true integration only comes when we hold management accountable and are able to measure results with inclusive and well-designed monitoring and evaluation systems.

While our gender policy articulates that accountability rests with senior management, in actuality there are still no real incentives or disincentives for senior managers in terms of promotion to ensure gender is being integrated into the mission's work. There are a lot of other mandates from Washington that supersede the goal on gender equality. We are supposed to report on some 'mandatory as applicable' indicators every year. However, many mission leaders see the 'as applicable' phrase, and the gender indicators just drop out.

Also, we still struggle to make sure that all our data are sex disaggregated, which became a requirement in the mid-1980s. We don't want colleagues to approach gender equality as just reaching that bare minimum. We do want people to think critically about these issues, but it's really difficult in a large bureaucracy where there's not a lot of investment in helping people think about these things. We only have one required on-line course that takes an hour. We do have other courses on gender integration, integrating approaches to gender-based violence into different development sectors (health, economic growth, DRG etc.) but these are not required, even for Agency gender advisors.

Our job as gender advisors is to help people understand why gender equality matters for their project outcomes – that is the lens that most colleagues use to measure their success. The rhetoric we use is that if we address gender inequality, you'll actually have better outcomes. People will ask, 'Well, what is the business case for investing in this field?' Usually, we have to position everything as economic arguments for why we should address gender inequality, for example, how it will increase GDP or business profits. And that works better than saying, 'Well, this is the feminist agenda or feminism is the right thing to do.' Gender equality is a human rights issue. It is very uncomfortable for us having to take this instrumentalist, neoliberal approach because it is a human rights issue more than anything but that's not how we can approach it.

Another dilemma is when Washington might fund research for three years on a politically driven topic like women's economic empowerment that is meant to inform our work in the field. But 'all politics are local' – so is gender. What worked in Uganda to 'empower women entrepreneurs' might not work in Lao PDR. USAID has a penchant to look for 'scalability'. That can present some problems for our gender work. For example, it is almost dogma that women's employment will lead to empowerment. Research emerging from Asia is indicating that if women earn enough to be able to bargain on decision-making in the home this might actually increase GBV. So, I would resist our Missions claiming that women's economic participation will empower women. Do women need to earn money? Absolutely, but we need to understand the dynamics in sub-contexts to ensure we are doing no harm.

Optimising spaces for change

Given the complexity and these challenges, we spend much of our time advocating for improvements or pointing out deficiencies; however, we try to never lose sight of the many successes we see in programming – it is always rewarding when a USAID staff member or implementing partner transforms into a gender and social inclusion champion in their work.

For example, in our regional work on natural resource management and the environment, USAID has made some major contributions, writ large. We have a couple of projects that had visionary people on the design team who insisted that gender be included. USAID's five-year project on fisheries in Southeast Asia has a 'pillar' on gender and decent work. That was designed before we arrived at our Mission. It is a highly technical project that has provided some really important insights into the work of men and women in the fisheries sector and has been a first for USAID for the level of detail and the attention to how gender dynamics vary between countries in this sector. Will's efforts have spurred our wildlife trafficking project to consider the different roles of men and women in wildlife trafficking and in consumer demand. Also, REDD+, a global initiative USAID participated in, did a gender analysis back in 2012 and the USAID manager of that project insisted the implementing partner hire a gender advisor. How much the findings and efforts of that project have been integrated by the many agencies implementing REDD+ is still a question.

Recently, there has been much more attention to gender and social inclusion at anything USAID/RDMA is designing because there is another relentless gender advisor there – Will. We had an amazing gender analysis done for a coffee project we are supporting. We learned so much: Kai was so excited about the project's potential to address real gender gaps as opposed to 'include women in training', which is a rather facile recommendation that is often seen. That is what a gender analysis should do! The data were persuasive in not just identifying a value chain that could be shaped more equitably, but in also identifying the concrete gender and social inclusion gaps that must be closed to realise the project's goals. We have had some gender analysis done for our new work in energy; it is still to be seen how well we at RDMA and gender advisor colleagues at the Missions can convince their colleagues that gender inequality is an important issue in the energy sector – it's more than turbines cranking out energy. The most important current challenge for us in our environment work is ensuring that activities to address gender inequality and prevent and respond to GBV and empower women and girls are integrated throughout our project activities.

Currently, we have GESI analyses underway to study basic education and literacy, disability, violent extremism, trade, natural resource management in Tibet, and digital connectivity and cybersecurity, among others, which will allow us to address gender gaps in multiple sectors across Asia.

There are promising opportunities to effect change and the commitment to realise them is unwavering despite thin resources and some corners of resistance. It remains to be seen, however, whether USAID will re-work its entire architecture and narrative of development in the future, responding to the vision of a more equitable and sustainable world.

References

Campbell, M. L., & Teghtsoonian, K. (2010). Aid effectiveness and women's empowerment: Practices of governance in the funding of international development. *Signs: Journal of Women in Culture and Society*, *36*(1), 177–202. https://doi.org/10.1086/652914

Cullen, P., & Murphy, M. P. (2018). Leading the debate for the business case for gender equality, perilous for whom?: Leading debate for the BCGE. *Gender, Work & Organization*, *25*(2), 110–126. https://doi.org/10.1111/gwao.12199

Eyben, R. (2015). Promoting gender equality in the changing global landscape on international development cooperation. In A. Coles, L. C. Gray, & J. H. Momsen (Eds.), *The Routledge Handbook of Gender and Development* (pp. 515–526). Routledge.

Ferguson, L. (2015). ''This Is Our Gender Person.'' *International Feminist Journal of Politics*, *17*(3), 380–397. https://doi.org/10.1080/14616742.2014.918787

Fraser, N. (2009). Feminism, capitalism and the cunning of history. *New Left Review*, *56*(2), 97–117.

Friedan, B. (1963). *The Feminine Mystique*. W.W. Norton & Company.

Girod, D. (2008). Cut from the same cloth? Bilateral vs multilateral aid. Annual Meeting of the International Political Economy Society, Philadelphia, Pennsylvania, USA.

Gulrajani, N., & Calleja, R. (2019). Understanding donor motivations: Developing the Principle Aid Index (Working Paper No. 548; p. 44). Overseas Development Institute (ODI).

Johnston, J., & Main, A. (2013). *Breaking Open the Black Box: Increasing Aid Transparency and Accountability in Haiti*. Center For Economic and Policy Research.

Kabeer, N. (2013). Gender equality and economic growth: Is there a win-win? (No. 417; IDS Working Paper). Institute for Development Studies. http://www.ids.ac.uk/publication/gender-equality-and-economic-growth-is-there-a-win-win

Mawdsley, E., Murray, W. E., Overton, J., Scheyvens, R., & Banks, G. (2018). Exporting stimulus and ''shared prosperity'': Reinventing foreign aid for a retroliberal era. *Development Policy Review*, *36*, O25–O43. https://doi.org/10.1111/dpr.12282

Murray, W. E., & Overton, J. (2016). Retroliberalism and the new aid regime of the 2010s. *Progress in Development Studies*, *16*(3), 244–260. https://doi.org/10.1177/1464993416641576

Norris, J. (2012, 18 July). Hired gun fight. *FP Foreign* Policy. https://foreignpolicy.com/2012/07/18/hired-gun-fight/#

Overholt, C., Anderson, M., Cloud, K., & Austin, J. E. (1985). *Gender Roles in Development Projects: A Case Book*. Kumarian Press.

Rathgeber, E. M. (1995). Gender and development in action. In M. H. Marchand & J. L. Parpart (Eds.), *Feminism/Postmodernism/Development* (pp. 2014–2220). Routledge.

Roberts, S. M. (2014). Development capital: USAID and the rise of development contractors. *Annals of the Association of American Geographers*, *104*(5), 1030–1051. https://doi.org/10.1080/00045608.2014.924749

Afterword

Gender expertise, environmental crisis and the ethos of care

We bring closure to this book at a very challenging moment. For many of us, the story of the COVID-19 pandemic slowly unfolded from the margins of our awareness, and with sheer disbelief we witnessed rapidly spiralling numbers of infection cases and deaths globally in a matter of weeks. It soon dawned on us that virtually everyone stood on the fault lines of its impact, and that this was going to radically change our lives. To break the chains of contagion, states ordered self-isolation, home quarantines, lockdowns, travel bans and border closures. Public health infrastructures responded through patchy regimes of testing, treatment and care for the ill, all carried out in isolation from loved ones. In many places, public health systems were put severely under stress with threadbare protection for many frontline health workers who themselves have fallen ill or have died. As the pandemic crisis unfolded, it became clear that we were not 'all in it together' as those in power claimed, but rather, the differential impacts of the virus and the political response were (and continue to be) marked by long histories of racial, class, gender, colonial and spatial inequalities (Aguirre, 2020).

We will forever remember 2020 as the period when coronavirus COVID-19 laid bare not only the decades-long weakening of public health systems in many countries but also, in an unprecedented way, the crisis of environment and development. Arundhati Roy (2020) wrote how the pandemic is a portal – a doorway to other possibilities and a chance to think the world anew. The question of who is doing this thinking, and what sorts of possibilities are finally realised is one for feminists in environment and development settings to take seriously. There is a political ecology to the pandemic, which requires an intersectional feminist analysis and response. We would like to situate this afterword in the current crisis gripping the world, with our gaze fixed on inequality and injustice in the time of pandemic and beyond, juxtaposed with the book's reflections on negotiating gender expertise.

Interconnectivities between society and nature

Reports and studies are now starting to emerge that highlight the effects of biodiversity loss, deforestation, extractive activities such as mining and massive road building, sale of live wild animals that can all inevitably spin off to create new viruses (Berkley, 2020; McMahon, 2020; Thomson Reuters Foundation, 2020; Vidal, 2020). 'We invade tropical forests and other wild landscapes, which harbour so many species of animals and plants – and within those creatures, so many unknown viruses. We cut the trees; we kill the animals or cage them and send them to markets. We disrupt ecosystems, and we shake viruses loose from their natural hosts. When that happens, they need a new host. Often, we are it' (Quammen, 2020). Protecting the dignity, land and human rights of indigenous peoples who occupy much of the earth's forested areas is also the best way to keep forests standing, which in turn reduces global warming and biodiversity loss, suggests one study (Walker et al., 2020). Continued marginalisation and exploitation of human and non-human groups also show us that pandemics not only produce differentiated and unequal effects, but that deep and widespread inequality may actually cause pandemics (Spinney, 2020).

Avoiding a 'shock doctrine'-style response to the ecological dimensions of pandemic means drawing on the feminist political ecology store cupboard, where lessons on working towards this in a just and collaborative way are well documented. As one example, the polycentric rooted networks approach detailed some time ago by Rocheleau and Roth (2007) has gained renewed significance. We see collaborations across different geographies enabling conversations and actions to address specific conjunctures of COVID-19 and its variegated impacts, brought about through intersecting inequalities of ecology, gender and class. Working in this way, recent posts from Maharashtra in India linking young academics and frontline activists in feminist political ecology question the dominance of patriarchal and market logics that have made food systems so vulnerable (Leonardelli et al., 2020; Still et al., 2020).

During this time of the pandemic, gender experts and researchers have exposed rising incidences of gender-based violence in locked-down communities as power dynamics are magnified behind the veil of privacy of home confinement. The abused are overwhelmingly women and children, and it is harder for victims to get away from perpetrators and access social protection services, such as violence and abuse shelters. Those working from home juggle between productivity pressures, children's tele-learning, and boredom, while others have lost daily waged jobs and are barely able to survive due to lack of unemployment protection especially in informal work sectors. The pandemic has shone a harsh light on the unequal workloads and economic hardships borne by different groups (Han & Resurrección, 2020; Lewis, 2020). With schools closed, the elderly sheltered, and the sick quarantined at home, the pandemic is increasing women's domestic and family workload and exposing them to greater health risks.

Under-valuing care has then become the most glaring truth about the coronavirus pandemic. From feminist political ecology/economy, we become starkly aware that the pandemic exposes how the feverish race for economic growth and production has severely impaired and compromised virtually all facets of social reproduction, including nature's reproduction. This lays bare the deep interconnections between society and nature when examined through a prism of care. The reduced public provision of care is demonstrated by the ever-growing incapacities of public health personnel to deal with the pandemic as it spreads. Public health care has been ghettoised and widely devalued in many so-called industrialised countries, no matter where it takes place and who performs it, the low pay often justified by representing such work as 'low-skilled' (Fraser, 2016; Razavi, 2011). Commercialised care, only affordable to some, highlights that the poorest are indeed the most uncared for. Paradoxically, as the world now teeters on the brink of a major catastrophe, it heavily and almost exclusively relies on available care in all its private and public forms for its survival in homes, hospitals, hospices, state quarantines, elderly care homes, and care shelters.

As the tectonic plates of the global economy shift in response to the pandemic, questions around care and social reproduction are being re-visibilised. From the vantage point of feminist political ecology, we see a re-politicisation of care and social reproduction with, on the one hand, political leaders strategising for a 'great reset' of capitalism (with a care economy to support that) (World Economic Forum, 2020) whilst others look for a more radical reset that enables a regenerative valuing of care against new forms of care extractivism (Bhattacharyya 2018; Harcourt 2020). Is gender expertise now crucial more than ever, and if so, what might this need to look like as the world is forced to confront deepening inequalities cross-cut by race and coloniality?

Gender expertise and epistemologies of care

Positioning gender expertise within the combined crises of care, environment and development is to push the agenda of another kind of epistemology that sets itself apart from surveillance and the management of nature and diseases. Tragically, care epistemologies have long been consigned to the backseat of technocracy and rational managerialism in environment and development institutions as many gender experts have expressed in this book's chapters. The almost daily invocation of science during the pandemic gives the impression that science is singular, uncontested and objective, operating outside politics and social norms (Leach, 2020; McGregor et al., 2020; Scoones, 2020). A call for a more pluralistic approach that attends to the social dynamics of the pandemic and its aftermath is required, learning from the prefigurative forms of solidarity, mutual aid and innovation that have emerged in diverse settings (Scoones, 2020). This compels us to re-visit some of the premises of gender expertise in this book about negotiating environment and development. Through reflexivity, we de-privilege forms of observation related to the 'objective and measurable' external world in the belief that it is possible to be independent from what is observed. To be reflexive is to acknowledge the social, experiential and emotional context that inevitably determine a person's interpretation of knowledge leading to how they use it to take action (Bradbury et al., 2019).

Gender expertise, especially where it is informed by intersectional feminism (Crenshaw, 1991) is explicitly sensitive to inequalities and thus necessarily seeks to democratise, ameliorate, care for and lend justice to situations where social exclusions have led to forms of disadvantage. Making care central to our work begins with a notion of the interconnectivity between people, nature and non-human beings or 'a social ontology of connection: foregrounding relationships of mutuality and trust', (Lawson, 2007: 4) that goes beyond defining care as uniquely and essentially a feminine or maternal trait, or solely the responsibility of women. Activists and academic feminist political ecologists are working from more pluralistic ontologies that extend and deepen analyses of care to explicitly re-include ecologies and the non-human in relations of care, establishing connections that market-based approaches to gender and the environment have erased.

The challenge remains 'to build a theoretical and empirical analysis of the structural and historical relationships producing disease, hunger, poverty, environmental decline and disasters – or more broadly – the need for care' (Lawson 2007: 8). As feminist researchers and activists are showing, an ethics of care brings into question prevailing 'principles of individualism, egalitarianism, universalism, and of society organized exclusively around principles of efficiency, competition, and a "right" price for everything' (Lawson, 2007: 8) and with this, a questioning of productivism and economic growth as the values driving our endeavours (Feminisms and Degrowth Alliance, 2020). The pandemic has brought to glaring view the failures of the combined and massive extraction from nature, labour and reproductive labour. With this, we come to understand that gender equality cannot be realised by separate and relatively isolated or piecemeal gender interventions or mainstreaming, but that policy-making institutions at different scales themselves need to be transformed (van Eerdewijk & Davids, 2013). Care not just as an object or arena for intervention, but care as a *verb* – describing the way we as feminists work together, ethically in the everyday. This means being able to bring a feminist ethics of care into mainstream environment and development practice in contexts where the logic of scientific 'progress' can discomfort these efforts.

We now turn to the critical efforts of gender experts in pressing for transformation.

Reflexivity as part of our work

The core undertaking in this book was to reflect on our work as gender experts, as we are primarily tasked to translate feminist knowledge in environment and development niches with an ambition to transform inequality, help put an end to all kinds of discrimination and injustice and save our planet. Far from being utopian, however, the revolution is a slow one, in many ways because as experts, power also works through us: thus, we are not separate from the hegemonic institutions we are trying to transform (Davids et al., 2014). This book has given us a space to pause and converse about self-reflexive understandings of gender expertise through many prisms, while simultaneously highlighting the complexities of practice (Prügl, 2016). Through our reflexivity, we also draw our attention to ways of building on the small margins of change, as radical efforts are not contingent on scale but depend on the depth of our engagement (Bacchi, 2009; Subrahmanian, 2004; van Eerdewijk & Davids, 2013). Out of these moments of reflection that punctuate our conversations in this book, we would like to nuance gender experts' negotiations with environment and development bureaucracies as they work across an epistemological interface with positivist science.

First, we reflected on how gender experts 'come into being' in science-led work environments. On a self-reflexive note, as we learned more about our contributors' journeys in becoming 'gender experts', we became increasingly aware that many of the questions and issues each has encountered have a long history, even as the terminologies and conceptual frameworks to name and analyse them evolve and unfold. We are reminded of our own early years as doctoral researchers in Indonesia and the Philippines studying complex subjectivities in upland communities. Becky, for instance, saw how the fluidity of gender subjectivities and divisions of labour left her 'unable to live up to the practical demands of natural science-based development research, which seems to need a stable, fixed and visible "object", not an "object" that is vested with agency' in a development project in Lampung, Indonesia (Elmhirst, 1998: 232). Likewise, Babette saw how ethnic minority women in upland Philippines gave little importance to civil society definitions of indigeneity in their market transactions whereas upland men vested indigeneity with important political capital: 'gender plays an important role in demonstrating that identities may shift, emerge or recede in particular situations of contingency as it is constantly in process' (Resurrección, 2006: 396). These recollections helped us further firm up our understanding of the situated and profoundly unstable nature of gender expertise as 'gender experts' themselves are continually being constituted by the contexts within which they work and wish to transform, in conjunction with their own truth commitments and personal-historical experiences. Maeve Nightingale (Chapter 3), Joy Clancy (Chapter 1), the SEI team (Natalia Biskupska, Marisa Escobar, Laura Forni, Emily Ghosh, Ha Nguyen, and Lisa Segnestam in Chapter 5) and Clara Mi Young Park (Chapter 4) in this book also highlighted how gender expertise is itself a site of contestation, often instantiating binary categories, where various truth claims, processes of personal awareness-raising, and diverse involvements in gender equality work can mark boundaries of inclusion and exclusion, as well as providing a community of practice with which to strike strategic alliances.

Second, we became increasingly aware of the intrinsically political nature of bringing in gender and social equity concerns into the epistemological orbits of environment and development organisations. Indeed, the entry and growing presence of gender expertise is ruffling the feathers of those who strictly adhere to the authority of positivist science in these realms. But as we hinted in the Introduction, gender experts' responses to resistance have been far from uniform, being contingent fundamentally on three issues. These are: (i) the distinct and perhaps hybrid ways with which their organisations frame the relationship between nature and society, that either possess the ambition to control environment and nature through technical and managerialist means or instead recognise that these are not externally induced but are created through social, economic, political and ecological interactions (Arora-Jonsson & Ågren, 2019); (ii) their own disciplinary backgrounds and truth commitments; and (iii) their own embodied locations and histories.

Some resist and carve out separate spaces such as Kate Lappin (Chapter 12) who disassociates her engagements from development technocracy as she has realised that alternative feminist fossil-free futures can be more fully realised through strong feminist social movements. We learned from Seema Kulkarni and Margreet Zwarteveen (Chapter 6) that many years of engagement and expert-led attempts to integrate social and gender inclusion in water management have led them to realise that the inclusion of gender expertise in programs stimulated by global water agreements may neutralise, instead of advance, a more political and feminist gender and water justice agenda. While they continue to politically engage with water authorities to advance this agenda, as experts, they will be more cautious about integrating gender with water management. Instead, they find it more fruitful to lay bare the plurality of fundamental goals and knowledges between technical water management experts and gender experts.

Others tactically blur the boundaries between gender expertise and the technical sciences in order to steer the boat in their direction by seeking complementarity and optimising co-learning opportunities. Nicoline de Haan, Carol Colfer and Marlène Elias of the CGIAR share the common interlocutory strategy of using grounded field-based experiences to open channels for engagement and dialogue.

The book also reveals that some gender experts skilfully tap and leverage institutional regimes that have been breeding grounds for gender and social inclusion efforts to grow over the years. In the process of translating feminism in technical settings, gender experts may partly subvert and partly comply with existing power relations by leveraging institutional mandates and capitalise on friendships such as the narratives shared by Annette Wallgren and Victor Tsang (Chapter 8). In Chapter 2, Gordon Prain tells us that epistemic parity between the social and technical sciences is persistently elusive but may be gained at certain temporal moments as gender, seniority and disciplinary background combine to render epistemic playing fields momentarily equal, an observation shared by Joy Clancy (Chapter 1).

Through strategic reflexivity, we also 'identify possible points of reversal or switches, whereby potential openings for struggle and contestation occur' (Bee & Sijapati Basnett, 2017: 797). We come to be aware that environment and development organisations are also sites of disciplining, sometimes remaining formidably skewed towards the institutional inertia of masculinised culture and practices. From the struggles and experiences of Hilde Jakobsen, Napapan van der Kinderen and Maria Holtsberg (Chapter 10) working in disaster risk reduction contexts, we gained valuable knowledge about which femininities were considered acceptable and the norms of masculinity and episteme that have become apparent by studying the social life of disaster and humanitarian organisations (Kronsell, 2005).

Third and finally, conversations with Sida (Chapter 13, Maria Von Berlekom, Eva Johansson, Orawan Raweekoon and AnnaKarin Norling) and USAID (Chapter 14, Kai Spratt and Charles 'Will' Lewis, Jr) confirm that gender equality is on top of most donor agendas but in uneven ways. This unevenness is in large part due to wider dynamics that involve conflating development with stimulating economic growth and globalising markets. They showed us that by being champions, donors have strategically catalysed gender equality efforts in their partner organisations but are sometimes distanced (and in some instances, powerless) from the internal resistances and knowledge/power tensions within these organisations that mediate and constrain the transformations that they aspire to. As such, they may in the end acquire positions of management and oversight rather than being active partners in the often-messy tasks of instigating change.

Conclusion: Staying with the trouble but staying on course

In their volume on investigating the politics of gender expertise and training, Bustelo et al. (2016) conclude that the more unequal the power (and knowledge dynamics) of the work or institutional context into which it is being introduced, it is significantly crucial to interrogate the ways with which gender is being integrated. We set out to examine what this might mean in technical-environmental 'research-for-development' settings where 'research' is dominated by the authority of natural science epistemologies and where 'development' is generally envisioned in the language of productivism and growth.

From our contributor discussions, we learn that gender experts, and indeed those that refuse this label, stay with the trouble despite seemingly insurmountable odds as they negotiate epistemic authorities in environment and development realms. What is heartening in our discussions is that a commitment to facilitating the birth of a better way and to a better world remains as a strong conviction. Rather than being derailed by the contradictions they face in their work, the gender experts who have contributed to this book have all, in their different ways, shown how in their work they draw on multiple sources, ranging from a previous academic training in feminist political ecology to solidarity with activists and others outside the mainstream development world. At various times, these provide 'ethical moorings', not for the purpose of marking professional standards or boundaries, but to ensure and uphold a transformative mission. Vigilance and reflexivity help resist hegemonising rationalities that depoliticise and technocratise the work of advancing gender equality in technical environments.

Circling back to the pandemic, and its portal-like qualities that offer not only a glimpse of what a future world might be, but an amplification of the inequalities and injustices that mark both the social and the ecological, such vigilance is required to serve a larger purpose. For now, the machineries of international development are likely to be deployed in whatever direction the 'great reset' is likely to take (World Economic Forum, 2020). Optimistically, responses to COVID-19 that have brought a rhetorical re-valuing of care, community-led solidarity and collective action and a glimpse of a more ecologically just world might reshape the norms and practices of environment and development. But this is not guaranteed. The threat of authoritarian populist governments bent on undemocratic hard-fisted technologies of rule that value or dispense with human and ecological life at the whim of market forces remains. More than ever before, *care-ful* feminist solidarities across activist, academic and bureaucratic spaces are required, recognising the role that each of us must take to maintain vigilance and counter any possibility that the opportunity for social and ecologically just transformation is lost.

References

Aguirre, M. (2020). The impact of COVID-19 is all down to inequality. *Open Democracy*, 7 April 2020. Retrieved from: https://www.opendemocracy.net/en/impact-covid-19-all-down-inequality/

Arora-Jonsson, S., & Ågren, M. (2019). Bringing diversity to nature: Politicizing gender, race and class in environmental organizations? *Environment and Planning E: Nature and Space*, *2*(4), 874–898. https://doi.org/10.1177/2514848619866381

Bacchi, C. (2009). The issue of intentionality in Frame Theory: The need for reflexive framing. In M. Verloo, P. Meier, & E. Lombardo (Eds.), *The Discursive Politics of Gender Equality: Stretching, Bending and Policy-Making* (pp. 30–41). Routledge.

Bee, B. A., & Sijapati Basnett, B. (2017). Engendering social and environmental safeguards in REDD+: Lessons from feminist and development research. *Third World Quarterly*, *38*(4), 787–804. https://doi.org/10.1080/01436597.2016.1191342

Berkley, S. (2020, 4 February). Novel Coronavirus is a reminder: The best defense against a new viral outbreak is early detection. *Scientific American Blog Network*. https://blogs.scientificamerican.com/observations/novel-coronavirus-is-a-reminder-the-best-defense-against-a-new-viral-outbreak-is-early-detection/

Bhattacharyya, G. (2018). *Rethinking Racial Capitalism: Questions of Reproduction and Survival*. London: Rowman & Littlefield.

Bradbury, H., Waddell, S., O' Brien, K., Apgar, M., Teehankee, B., & Fazey, I. (2019). A call to action research for transformations: The times demand it. *Action Research*, *17*(1), 3–10. https://doi.org/10.1177/1476750319829633

Bustelo, M., Ferguson, L., & Forest, M. (Eds.). (2016). *The Politics of Feminist Knowledge Transfer: Gender Training and Gender Expertise*. Palgrave Macmillan.

Crenshaw, K. (1991). Mapping the margins: Intersectionality, identity politics, and violence against women of color. *Stanford Law Review*, 1241–1299.

Davids, T., van Driel, F., & Parren, F. (2014). Feminist change revisited: Gender mainstreaming as slow revolution. *Journal of International Development*, *26*(3), 396–408. https://doi.org/10.1002/jid.2945

Elmhirst, R. (1998). Reconciling feminist theory and gendered resource management in Indonesia. *Area*, *30*(3), 225–235. https://doi.org/10.1111/j.1475-4762.1998.tb00067.x

Feminisms and Degrowth Alliance (FaDa) (2020) *Collaborative Reflections on the COVID-19 Pandemic and the Politics of Social Reproduction*. 20 April 2020. https://www.degrowth.info/en/2020/04/feminist-degrowth-collaborative-fada-reflections-on-the-covid-19-pandemic-and-the-politics-of-social-reproduction/

Fraser, N. (2016). Contradictions of capital and care. *New Left Review*, 100 https://newleftreview.org/issues/II100/articles/nancy-fraser-contradictions-of-capital-and-care

Han, J. Y.-C., & Resurrección, B. P. (2020, 6 April). *Covid-19 is also a Crisis of Care*. https://www.sei.org/perspectives/covid-19-exacerbates-social-and-gender-inequalities/

Harcourt, W. (2020). *Regenerative Practices of Care During and beyond the Pandemic.* WEGO-ITN blogpost 21 May 2020. https://www.wegoitn.org/regenerative-practices-of-care-during-and-beyond-the-pandemic/

Kronsell, A. (2005). Gendered practices in institutions of hegemonic masculinity: Reflections from feminist standpoint theory. *International Feminist Journal of Politics, 7*(2), 280–298. https://doi.org/10.1080/1461674050006517О

Lawson, V. (2007). Geographies of care and responsibility. *Annals of the Association of American Geographers, 97*(1), 1–11. https://doi.org/10.1111/j.1467-8306.2007.00520.x

Leach, M. (2020). *Development must Change in the Face of Injustice and Inequality.* Institute of Development Studies. https://www.ids.ac.uk/opinions/development-must-change-in-the-face-of-injustice-and-inequality/

Leonardelli, I. Tozzi, A., Still, E., & Kumar, S. in conversation with Seema Kulkarni and Sheha Bhat. (2020). *A Pandemic of Blindness: Uneven Experiences of Rural Communities under COVID-19 Lockdown in India, Part 2.* Undisciplined Environments blog post published 30 April 2020. https://undisciplinedenvironments.org/2020/05/05/a-pandemic-of-blindness-uneven-experiences-of-rural-communities-under-covid-19-lockdown-in-india-part-ii/

Lewis, H. (2020, 20 March). The Coronavirus is a disaster for feminism. *The Atlantic.* https://www.theatlantic.com/international/archive/2020/03/feminism-womens-rights-coronavirus-covid19/608302/

McGregor, H., Wilkinson, A., Parker, M., & Leach, M. (2020). *COVID-19: A Social Phenomenon Requiring Diverse Expertise.* Brighton, UK: Institute of Development Studies. Published 20 March 2020. https://www.ids.ac.uk/opinions/covid-19-a-social-phenomenon-requiring-diverse-expertise/

McMahon, J. (2020, 21 March). How deforestation drives the emergence of novel Coronaviruses. *Forbes.* https://www.forbes.com/sites/jeffmcmahon/2020/03/21/how-deforestation-is-driving-the-emergence-of-novel-coronaviruses/#1c784f1d1723

Prügl, E. (2016). How to wield feminist power. In M. Bustelo, L. Ferguson, & M. Forest (Eds.), *The Politics of Feminist Knowledge Transfer: Gender Training and Gender Expertise* (pp. 25–42). Palgrave Macmillan.

Quammen, D. (2020, 28 January). We made the Coronavirus epidemic. *The New York Times.* https://www.nytimes.com/2020/01/28/opinion/coronavirus-china.html

Razavi, S. (2011). Rethinking care in a development context: An introduction. *Development and Change, 42*(4), 873–903. https://doi.org/10.1111/j.1467-7660.2011.01722.x

Resurrección, B. P. (2006). Gender, identity and agency in Philippine upland development. *Development and Change, 37*(2), 375–400.

Rocheleau, D., & Roth, R. (2007). Rooted networks, relational webs and powers of connection: Rethinking human and political ecologies. *Geoforum, 38*(3), 433–438.

Roy, A. (2020). The pandemic is a portal. 3 April, 2020. Financial Times, London. https://www.ft.com/content/10d8f5e8-74eb-11ea-95fe-fcd274e920ca

Scoones, I. (2020). Science, uncertainty and the COVID-19 response. Published 16 March 2020. Brighton, UK: Institute of Development Studies. https://www.ids.ac.uk/opinions/science-uncertainty-and-the-covid-19-response/

Spinney, L. (2020, 12 April). Inequality doesn't just make pandemics worse – it could cause them. *The Guardian*, 5.

Still, E., Kumar, S., Leonardelli, I., & Tozzi, A. (2020). A pandemic of blindness: Uneven experiences of rural communities under COVID-19 lockdown in India, Part 1.

Undisciplined Environments blog post published 30 April 2020. https://undisciplinedenvironments.org/2020/04/30/a-pandemic-of-blindness-uneven-experiences-of-rural-communities-under-covid-19-lockdown-in-india-part-i/

Subrahmanian, R. (2004). Making sense of gender in shifting institutional contexts: Some reflections on gender mainstreaming. https://doi.org/10.1111/j.1759-5436.2004.tb00160.x

Thomson Reuters Foundation. (2020, March 11). Asia's rapid urbanisation, deforestation linked to deadly viruses. https://www.bangkokpost.com/world/1876459/asias-rapid-urban...AR2-fc0_DepiF4WHt3wi4vFZc712Am3EpGT5DNYrXqpqfyYD8NGKPEPmZGs

van Eerdewijk, A., & Davids, T. (2013). Escaping the mythical beast: Gender mainstreaming reconceptualised. *Journal of International Development*, *26*(3), 303–316. https://doi.org/10.1002/jid.2947

Vidal, J. (2020, 18 March). 'Tip of the iceberg': Is our destruction of nature responsible for Covid-19? *The Guardian*. https://www.theguardian.com/environment/2020/mar/18/tip-of-the-iceberg-is-our-destruction-of-nature-responsible-for-covid-19-aoe?

Walker, W. S., Gorelik, S. R., Baccini, A., Aragon-Osejo, J. L., Josse, C., Meyer, C., Macedo, M. N., Augusto, C., Rios, S., Katan, T., Souza, A. A. de, Cuellar, S., Llanos, A., Zager, I., Mirabal, G. D., Solvik, K. K., Farina, M. K., Moutinho, P., & Schwartzman, S. (2020). The role of forest conversion, degradation, and disturbance in the carbon dynamics of Amazon indigenous territories and protected areas. *Proceedings of the National Academy of Sciences*, *117*(6), 3015–3025. https://doi.org/10.1073/pnas.1913321117

World Economic Forum (WEF) (2020) The Great Reset. https://www.weforum.org/great-reset/

Index

Please note that references to Notes are indicated by the letter 'n' followed by the Note number.